Software Adaptation in an Open Environment

A Software Architecture Perspective

Software Adaptation in an Open Environment

A Software Architecture Perspective

Yu Zhou and Taolue Chen

CRC Press
Taylor & Francis Group
Boca Raton London New York

CRC Press is an imprint of the
Taylor & Francis Group, an **informa** business

AN AUERBACH BOOK

CRC Press
Taylor & Francis Group
6000 Broken Sound Parkway NW, Suite 300
Boca Raton, FL 33487-2742

First issued in paperback 2020

ISBN-13: 978-1-138-74347-2 (hbk)
ISBN-13: 978-0-367-65803-8 (pbk)

This book contains information obtained from authentic and highly regarded sources. Reasonable efforts have been made to publish reliable data and information, but the author and publisher cannot assume responsibility for the validity of all materials or the consequences of their use. The authors and publishers have attempted to trace the copyright holders of all material reproduced in this publication and apologize to copyright holders if permission to publish in this form has not been obtained. If any copyright material has not been acknowledged please write and let us know so we may rectify in any future reprint.

Visit the Taylor & Francis Web site at
http://www.taylorandfrancis.com

and the CRC Press Web site at
http://www.crcpress.com

Contents

Foreword

Software is eating the world whilst the world is eating more and more software. The interplay between software and the world drives the evolution of software development methodologies. The way of constructing software is undergoing a fundamental paradigm shift. The execution environment of modern software is becoming more open, dynamic, and volatile. Such openness brings grand challenges to the adaptability of the inhabitant software systems. Since continuously delivering software satisfying users' needs is always a timeless pursuit of software developers, self-adaptation attracts considerable attention from both industry and academia. There has been a lot of research conducted in this area, inspired by applications from a multitude of disciplines. This can be evidenced by a proliferation of new adaptation techniques and frameworks that have emerged in recent years.

Among others, software architecture related techniques represent an important subject. This is partly due to the increasing application of component based systems in the open environment, such as web services. Software architecture provides an adequate abstraction level, as well as an effective way to guide the adaptation.

This book is one of the first monographs to address software adaptation in an open environment from a software architecture perspective. The two authors are active researchers have great experience in this field and provide a comprehensive discussion of current adaptation frameworks in the light of software architecture. This includes service discovery and interaction adaptation, adaptive component migration, or context and ontological models. The chapter on formal modeling covers the essence of adaptation rules, conflict detection, or verification of dynamic evolution.

The book aims both at practitioners and researchers and conveys the foundations of software adaptation. It describes verification techniques and frameworks and as such provides an excellent reference to this domain of dynamic software system evolution.

Zurich, December 12, 2016

Harald Gall

Preface

With the rapid development of computing and network technology, the operating environment of modern software systems is becoming increasingly open, dynamic, and uncontrollable. New computing paradigms, such as pervasive computing and cloud computing, are emerging. Among these paradigms, the common features of the environment exhibit an ever-growing trend of openness. This trend has a significant impact on software development and interaction. Moreover, a large number of systems are composed of distributed and autonomous components. The high reliance on software requires the system's robustness, *i.e.*, continual availability and satisfactory service quality. This requirement gives rise to the popularity of research on the self-adaptive software in such an open environment.

Traditional software adaptation approaches usually simplify the context and select some key variables from a specific application domain, and the adaptation logic is predefined, hard-wired, and mixed with business logic. Such an approach has the advantage of quick response. Nevertheless, it supports low reusability and adaptability in heterogeneous situations. Thus, in an open environment which is characterized by its dynamism and heterogeneity, software adaptation research faces many new challenges. The context is more complex than before. Diversity of platforms, variety of user preferences, difference of service discovery, and interaction protocols all possibly affect the system's adaptive behaviors. Therefore, it requires elaborate inspection of the context features and explicit modeling techniques. By incorporating its semantic information, the context changes can be understood by the software applications, and the adaptive behavior is conducted accordingly. However, the richness and complexity of context information in an open environment will also bring about other problems. To name a few, multiple adaptation rules are possibly activated concurrently as these rules are not necessarily orthogonal. The problem is how to detect the potential conflict and dependency relations. Aside from the adaptation enabling techniques, the problem is how to assure the dynamic evolution process is consistent with the specification. In the heterogeneous protocol environments, the problem is how to enable adaptive service discovery and interaction so as to provide continual service availability. In the case of mobility, the problem is how to support adaptive component-level migration to improve users' satisfaction and reduce unnecessary overhead. Traditional solutions originated from a relatively closed environment with limited applicability and flexibility.

For component based software systems running in an open environment, such as the Internet, despite the variety of the underlying implementation details, they share a common set of characteristics: high-level separation of computation and coordination, loosely coupling and high autonomous entities, protocol-support interactions, etc. These characteristics can be well captured by the concept of software architecture which focuses on the abstract view of constituent entities, their interactions, patterns of composition, and global constraints. In consideration of the increasing openness and autonomy of these distributed components, attention has gradually shifted from components' internal details to the composition and coordination of these components. In these situations, software architecture offers an adequate level of granularity and becomes a crucial artifact for the research on self-adaptation in software engineering. Having observed this, this book attempts to address the aforementioned problems from a perspective of software architecture and presents our recent research on the efforts of engineering self-adaptive software systems.

Constructing self-adaptive software is intriguing but not easy. It involves knowledge and expertise from multiple disciplines. This book does not aim to provide a comprehensive encyclopedia of building self-adaptive software. Instead, it focuses on the challenges raised by the open environment and is intended to be used as a general introduction to the engineering of self-adaptive software in such an environment.

Acknowledgements.

The authors would like to thank Professor Jian Lü (Nanjing University), Professor Harald Gall (University of Zurich), Professor Xiaoxing Ma (Nanjing University), Professor Luciano Baresi (Politecnico di Milano), Professor Carlo Ghezzi (Politecnico di Milano), Professor Jiannong Cao (Hongkong Polytechnical University), Professor Zhiqiu Huang (Nanjing University of Aeronautics and Astronautics), Professor Xianping Tao (Nanjing University), Dr. Andrea Mocci (University of Lugano), Professor David S. Rosenblum (National University of Singapore), Professor P. S. Thiagarajan (National University of Singapore), Professor Mingsheng Ying (University of Technology Sydney), Professor Yuan Feng (University of Technology Sydney), Professor Marta Kwiatkowska (University of Oxford), Dr. Guoxin Su (National University of Singapore), and Dr. Tingting Han (Birkbeck, University of London) for their collaboration, support, and guidance of the work.

Yu Zhou is partially supported by the Natural Science Foundation of Jiangsu Province under grant No. BK20151476, the National Basic Research Program of China (973 Program) under grant No. 2014CB744903, the National High-Tech Research and Development Program of China (863 Program) under grant No. 2015AA015303, the Collaborative Innovation Center of Novel Software Technology and Industrialization, and the Fundamental Research Funds for the Central Universities under grant No. NS2016093. Taolue Chen is partially supported by EPSRC grant (EP/P00430X/1), European CHIST-ERA project SUCCESS, ARC Discovery Project (DP160101652), Singapore MoE AcRF Tier 2 grant (MOE2015-T2-1-137), NSFC grant (No. 61662035), and an overseas grant from the State Key Laboratory of Novel Software Technology, Nanjing Unviersity (KFKT2014A14).

Authors

Dr. Yu Zhou is an associate professor at the College of Computer Science and Technology at Nanjing University of Aeronautics and Astronautics (NUAA), China. He received a doctoral degree in computer science from the Nanjing University, China in 2009. From 2010 to 2011, he conducted postdoctoral research on software engineering at Politechnico di Milano, Italy. From 2015 to 2016, he visited SEAL lab at the University of Zurich on sabbatical, where he is also an adjunct researcher. He is currently a senior member of the China Computer Federation (CCF) and a member of the Technical Committee on System Software of CCF. He has broad interests in software engineering with a focus on software evolution and reliability analysis.

Dr. Taolue Chen received bachelor's and master's degrees from the Nanjing University, China, both in computer science. He was a junior researcher (OiO) at CWI and acquired a PhD degree from the Free University Amsterdam, The Netherlands. He is currently a senior lecturer at the Department of Computer Science, Middlesex University London, UK. Prior to this, he was a research assistant at the University of Oxford, UK, and a postdoctoral researcher at the University of Twente, The Netherlands. His research interests include formal verification and synthesis of stochastic systems, model checking, concurrency theory, process algebra, and computational complexity.

List of Figures

List of Tables

BASICS AND FRAMEWORK

I

Chapter 1

Introduction

CONTENTS

It has long been a consensus in software engineering that software entity is constantly subject to pressure for change [30]. Numerous research efforts from various aspects have been devoted to addressing the issue. This can be observed from the perspective of the evolution of software process models and methodology. The waterfall process model first came to debut in the early 1970s. However, its assumption that each step could be done perfectly before moving to the next step is not true in many real projects, mostly due to the changes introduced from the requirements and environments. This fact had led to other iterative models such as incremental and spiral models, and the more recent agile models and DevOps models [19]. In accord with that, software methodology also evolves gradually from structure-oriented, then object-oriented, later component-oriented, to nowadays service-oriented. Each stage supports a more coarse-grained and loosely-coupled programming style than its predecessor. Correspondingly, the research focus shifts steadily from *programming-in-the-small* to *programming-in-the-large* [54].

These efforts have undoubtedly enhanced software robustness and productivity greatly. However, it is undeniable that the problem itself evolves as well. Remarkably, the operating environment of modern software is not as isolated or closed as before. In particular, with the rapid development of computing technology, the Internet, which was merely a portal of data and resources at its inception, has become the largest platform for software applications. As outlined in a vision document by the European Commission, *Internet-of-Contents and Knowledge*, *Internet-of-Things*, and *Internet-of-Services* are the three pillars of future Internet [138]. Different from traditional computing paradigms, the Internet is characterized by its scalability, openness, and heterogeneity. More concretely, it has the following salient features: decentralized distribution, highly autonomous network nodes, heterogeneous devices, unpredictable entity's behavior, potential security threats, personalized usage, and co-existence of various network protocols [187]. Many factors introduced by this kind of open environment, such as different types of platforms, variable conditions of hardware, and heterogeneous networks and protocols, have a significant impact on the software artifacts running on top of the Internet. The underlying characteristics of such an environment pose strict requirements on the *adaptability* of the inhabitant software systems. The mutual interplay between the environment and the underlying application gives birth to new software paradigms, such as *internetware*, which largely denotes a class of systems characterized by flexibly evolvable, continually reactive, and multiple objectives oriented in the open environment [120, 187].

Adaptability is an important dimension of the software quality attributes. In IEEE standard glossary of software engineering terminology [27], adaptability is defined as "The ease with which a system or component can be modified for use in applications or environments other than those for which it was specifically designed." Clearly, since the very beginning, building robust software that can adapt in the presence of adverse conditions has always been one of the focuses of software engineering. Numerous research efforts from different dimensions have been invested to address this issue. In almost every stage of the software development life cycle, such as abstraction, specification, implementation, and maintenance, we can identify these efforts. Indeed, we have witnessed the evident progress of enhancing adaptability in software design methodology and programming language mechanisms, such as design patterns, polymorphism, dynamic binding, etc. In design patterns, one of the underpinning principles is *open/closed principle* which states that software modules should be open for extension, but closed for modification. Meanwhile, polymorphism and dynamic binding allow that the overridden method invocation can be decided at runtime.

However, the extant approaches largely originate from the traditional environment with the "closed environment" assumption. Changes are anticipated and hard-wired in the source code. It works well as long as that assumption holds. However, in an open environment, at design time, developers cannot anticipate and plan all possibilities before the code delivery. Therefore, in case of the occurrence of unanticipated events at runtime, the software has to be shut-down and maintained manually so as to assure it operates properly. Needless to say, software is expected to continue running in case of such events, since nowadays it is eating the world and plays more and more important roles in business areas, critical missions, and our everyday life. The cost of shutting down the applications and evolving them off-line is surprisingly high. For example, a technology market research firm, IHS Inc., published a survey on *"the cost of server, application and network downtime"* in January 2016. The result reveals that the cost of each downtime event ranges from one million dollars for a typical middle-sized company to more than 60 million for a large company. The total downtime cost for North American businesses can sum up to 700 billion dollars[1]. This calls for the ability to be *self-adaptive* for modern software systems which remains a great challenge today. There is still a huge gap between fact and expectation. Throughout recent years, an ever-growing research effort is devoted to engineering such systems, both in academia and industry.

Traditional self-adaptive systems, such as robotics, avionics, and vehicle control systems, usually leverage some domain-specific algorithms or exception handling mechanisms to accommodate variabilities. As a result, this solution lacks generality and supports poor reusability. Supporting self-adaptation in a dynamic and open environment is far from a trivial task. Many efforts from multiple disciplines have contributed to this subject [158]. From the dimension of artificial intelligence and knowledge engineering, new dynamic planning or machine learning-based algorithms can be developed to guide the adaptation process. From the dimension of control theory, some parameter-tuning techniques based on specific control models have been proposed. From the dimension of

---

[1] http://press.ihs.com/press-release/technology/businesses-losing-700-billion-year-it-downtime-says-ihs

software engineering, many research areas are related, such as service computing, component technology, and software architecture [52].

Recently, the dynamic software architecture-based adaptation [75, 110] is gaining increased popularity due to its unique advantages. As software architecture is the global abstraction of the system and embeds the design decisions, it reflects the essential view of the application which is very useful to cope with the growing complexity of modern software. Secondly, nowadays, more and more software applications are developed through the composition from elemental autonomous components. The structure of such composition is not as static as before, and the elements can dynamically join and leave the integrated applications. Therefore, the composition and reconfiguration of the application structure can be naturally modeled by the software architecture dynamism. Moreover, decades of research on software architecture have resulted in a number of models and techniques which provide a useful knowledge base to analyze adaptation-related properties.

Given the above considerations, there is growing recognition that software architecture reconfiguration can be an effective way to enable software adaptation with the adequate generality and granularity. However, architectural reconfiguration is only the *action* part of adaptation. In an open environment such as the Internet, there are many other concerns to be taken into account during the adaptation process. This book focuses on the issues and attempts to give systematic solutions, instead of ad hoc ones, from a perspective of software architecture. The book inspects the characteristics of the open environment, and addresses the problems such as context modeling and reasoning, adaptive component-level migration, adaptive service discovery and interaction with heterogeneous protocols, conflicts/dependency detection for multiple adaptation rules, consistency verification, and adaption decision-making process.

This chapter is intended to be a general introduction to the research on software self-adaptation. Some new computing paradigms and methodologies emerged in open environments are introduced as the background in Section 1.1. Then we proceed to explain some basic concepts of self-adaptation in Section 1.2 and Section 1.3 respectively. The problems of self-adaptation in an open environment are given in Section 1.4 followed by an overview of the book organization in Section 1.5.

1.1 Novel Computing Paradigms and Methodologies

In this section, we will review some representative computing paradigms and methodologies that have emerged recently and are relevant with adaptation in an open and dynamic environment.

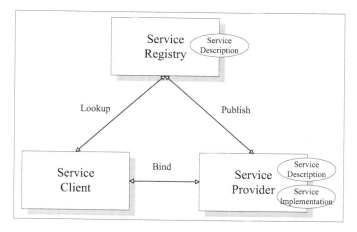

Figure 1.1: Basic model of SOA

Service computing and architecture

Service computing utilizes services as basic components for developing applications. Services are self-describing, platform-agnostic computational elements that support rapid, low-cost composition of distributed applications. Services perform functions, which can be anything from simple requests to complicated business processes. As on the Internet, it is impossible to require all the communication parties use the same platform and language. Therefore, services allow organizations to exhibit their core competencies programmatically over the Internet (or intra-net) by using standard (XML-based) languages and protocols, and being implemented via a self-describing interface based on open standards [140].

To facilitate this process, service oriented architecture (SOA) is proposed. The basic SOA model has three elements, *i.e.*, service provider, service registry, and service requestor. Service provider first needs to publish their service descriptions in the service registry. Service requestors first look up the service information at the registry center. After binding the corresponding information, the requestors will interact with the provider using some standard interfaces. This process is described in Figure 1.1.

In [100], Huhns and Singh summarize some key features of service oriented computing as below:

- Loose coupling. Tight transactional properties generally do not apply among components because conventional software architectures do not typically include transactional managers. Some high-level contractual relationships that specify component interactions to achieve system-level consistency should be considered.

- Implementation neutrality. The interface for each component matters most, because we cannot depend on the interacting components' implementation details, which can be unique. In particular, a service-based approach cannot be specific to a set of programming languages, which cuts into the freedom of different implementers and rules out the inclusion of most legacy applications.

- Flexible configurability. An SOA system is configured late and flexibly, which means that different components are bound to each other late in the process. Thus, the configuration can change dynamically as needed without loss of correctness.

- Persistence. Services do not necessarily require a long lifetime. Because we are dealing with computations among autonomous heterogeneous parties in a dynamic environment, we must always be able to handle exceptions. The services must exist long enough to detect any relevant exceptions.

- Granularity. An SOA's participants should be modeled and understood at a coarse granularity. Instead of modeling interactions at a detailed level, the high-level qualities that are visible for business contracts among the participants should be captured.

- Teams. Rather than framing computations centrally, we should think in terms of how autonomous parties, working on a team as business partners, realize those computations.

Because of the characteristics above, service based systems allow more space for adaptation. Several techniques, such as orchestration and choreography are proposed to describe aspects of creating business processes from composite web services. Orchestration refers to an executable business process that can interact with both internal and external web services and the interactions occur at the message level. Choreography tracks the message sequences among multiple parties and sources, rather than a specific business process that a single party executes [142]. Related specification standards are proposed, such as BPEL [72], WSCI [8], to describe this process. These techniques focus on the weaving script rather than on the internal business-level programming logic. Due to the separation of concerns, the adaptation of such systems requires the rewriting or recomposing of these specifications.

Autonomic computing

In March, 2001, IBM announced the initiative of *autonomic computing*. The idea was inspired by human's autonomic nervous system which can manage the body conditions without explicit intervention. Generally, the autonomous systems have the following characteristics [74]:

- To be autonomic, a system needs to "know itself" and consist of components that also possess a system identity.

- An autonomic system must configure and reconfigure itself under varying and unpredictable conditions.

- An autonomic system never settles for the status quo - it always looks for ways to optimize its workings.

- An autonomic system must perform something akin to healing - it must be able to recover from routine and extraordinary events that might cause some parts to malfunction.

- A virtual world is no less dangerous than the physical one, so an autonomic computing system must be an expert in self-protection.

- An autonomic computing system knows its environment and the context surrounding its activity and acts accordingly.

- An autonomic system cannot exist in a hermetic environment (and must adhere to open standards).

- Perhaps most critical for the user, an autonomic computing system will anticipate the optimized resources needed to meet a user's information needs while keeping its complexity hidden.

In short, the above characteristics can be summarized into four fundamental categories, *i.e.*, *self-configuring*, *self-healing*, *self-optimizing*, and *self-protecting*.

Self-configuring means that the autonomic systems can configure themselves automatically in accordance with high-level policies representing business-level objectives. For complex systems, manual configuration is time-consuming and error-prone, and thus self-configuration is desirable in these situations. *Self-healing* means that autonomic computing systems can detect, diagnose, and repair localized problems resulting from bugs or failures in software or hardware. A specific problem-diagnosis component is needed to analyze information from the context and to respond accordingly. *Self-optimizing* means that autonomic systems continually seek ways to improve their operation by identifying and seizing opportunities to make themselves more efficient in performance or cost. The components will proactively seek to upgrade their function by finding, verifying, and applying the latest updates. *Self-protecting* means that autonomic systems defend themselves as a whole against large-scale, correlated problems arising from malicious attacks or cascading failures that remain uncorrected by self-healing measures. It also means that the system can anticipate problems based on early reports from sensors and take steps to avoid or mitigate them [106].

IBM has suggested a MAPE-K (Monitor, Analyze, Plan, Execute, Knowledge) reference control loop model for autonomic computing [101] which is depicted by Figure 1.2. The model is used to describe the autonomous component.

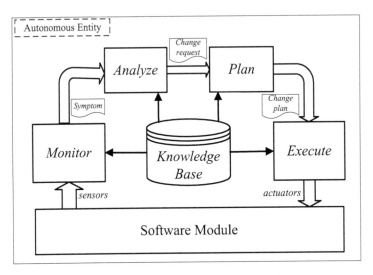

Figure 1.2: MAPE-K reference model

In this sense, the autonomous component is a special kind of intelligent agent situated in a context. The system is composed by a set of such autonomous entities. In terms of autonomy, each entity further consists of utility components to monitor the running environment, to analyze the change necessity, to plan the change actions, and to execute the changes. *Monitor* collects runtime information from physical/logical sensors and generates symptom reports to the *Analyzer* component which in turn generates a change request to the *Plan* component. Then the *Plan* component decides a change plan and effects the object software module accordingly through actuators. This cycle is an intelligent process based on a knowledge database.

To fully realize the vision of autonomic computing, grand challenges remain. Despite the difficulties, there are some research projects, such as ABLE toolkit [24] and autonomic toolkit [102]. Huhns *et al.* proposed a multi-agent based approach to enhance software's adaptability in the presence of adverse situations [99]. These projects have demonstrated the feasibility of the basic ideas in autonomic computing.

Grid computing

Grid computing originates from the fast growth of the Internet and the availability of powerful computers, as a particular form of distributed computing. Different from traditional paradigms, it views the network of computers as a single, unified computing resource. The purpose of a grid computing project is usually to solve a single but complicated problem which usually involves huge computations. By the strategy of divide-and-conquer, the problem is cut into smaller tasks and located to member computers within the grid. To achieve the goal, grid computing attempts to cluster or couple a wide variety of resources including supercomputers, storage systems, data sources, and special classes of geographically distributed devices and use them as a single unified resource, thus forming what is popularly known as a "computational grid" [11].

Due to the characteristics of the Internet, the resources for grid applications are not limited to a single site or local network. Instead, it can bind resources globally. In this light, grids enable users to solve large or new problems by utilizing the available resources together. The characteristics of computational grids are listed as below [11]:

- Heterogeneity. A grid involves a multiplicity of resources that are heterogeneous in nature and might span numerous administrative domains across wide geographical distances.

- Scalability. A grid might grow from a few resources to millions.

- Dynamism or adaptability. In a grid, a resource failure is the rule, not the exception. With so many resources in a grid, the probability of some resource failures is inevitably high.

Owing to the above characteristics, the grid computing paradigm is suitable for those applications with extremely high demands for data manipulation and calculation. The concept was initiated as a project to link super-computing sites. Now, there are many applications benefiting from the grid infrastructure. Multiple grid projects have been started. The most famous grid project is the Worldwide Large Hardron Collider (LHC) Computing Grid (WLCG) [163] which is coordinated by CERN [2]. The interconnection among many associated national and international grids across the world, such as European Grid and Open Science Grid makes WLCG the world largest computing grid. Some other notable cases are: SETI@Home [6], ACQUA@Home [104], and DAS [12].

Cloud computing

Cloud computing is the next natural step in the evolution of on-demand information technology services and products [180]. Similar to grid computing, cloud computing also attempts to abstract and virtualize resources on the Internet. But different from grid computing, it is not oriented to huge computing tasks. Instead, it is mainly designed for delivering IT services as computing utilities.

Under the umbrella of cloud computing, it can use a storage cloud to hold application, business, and personal data. Or it can be the ability to use a handful of web services to integrate photos, maps, and GPS information to create a mashup in customer Web browsers [1]. It can also be the ability to use applications on the Internet that store and protect data while providing a service. In [9], Armbrust *et*

---

[2]CERN is the acronym for European Organization for Nuclear Research situated in Switzerland.

al. believe that cloud computing refers to both the applications delivered as services over the Internet and the hardware and systems software in the data-centers that provide those services. In this sense, the data-center hardware and software is what they call a *cloud.*

In [71], Foster *et al.* define cloud computing as a large-scale distributed computing paradigm that is driven by economies of scale, in which a pool of abstracted, virtualized, dynamically-scalable, managed computing power, storage, platforms, and services are delivered on demand to external customers over the Internet. A more widely accepted definition is given by the National Institute of Standards and Technology (NIST), *i.e.,* cloud computing is a model for enabling ubiquitous, convenient, on-demand network access to a shared pool of configurable computing resources (*e.g.,* networks, servers, storage, applications, and services) that can be rapidly provisioned and released with minimal management effort or service provider interaction [128]. Three service models coexist in cloud computing, *i.e.,* software as a service (SaaS), platform as a service (PaaS), and Infrastructure-as-a-service (IaaS). Briefly speaking, SaaS denotes the services provided to customers to use provider-owned applications through the network. PaaS denotes the services provided to customers to develop, run, and maintain their own applications through the network. IaaS denotes the whole infrastructure, such as virtual machines, including application servers, storage, and other fundamental computing resources are provided to customers.

Although the elements of cloud computing, such as virtualization, SaaS, and the Internet, are not novel concepts individually, their combination brings some new highlights. In [9], Armbrust *et al.* summarized the following three new aspects:

1. The illusion of infinite computing resources available on demand, thereby eliminating the need for clouding computing users to plan far ahead for provisioning;

2. The elimination of an up-front commitment by cloud users, thereby allowing companies to start small and increase hardware resources only when there is an increase in their needs;

3. The ability to pay for use of computing resources on a short-term basis as needed and release them as needed, thereby rewarding conservation by letting machines and storage go when they are no longer useful.

In the early ages of cloud computing, a few examples in practice have been designed to suggest likely directions. In [92], Hayes enumerates four representative categories.

1. *Wordstar for the web. Google Docs* is such a case in point. The set of programs includes a word processor, a spreadsheet, a form, and a presentation tool.

2. *Enterprise computing in the cloud*. Maybe the most famous example of this kind is the *salesforge.com*. Different from traditional standalone Customer Relationship Management (CRM) solutions, it offers a suite of programs over the Internet. The programs can be flexibly tailored according to the customers' requirements.

3. *Cloudy infrastructure*. Amazon Elastic Computing Cloud (EC2)[3] is a web service that provides resizable compute capacity in the cloud and allows to quickly scale capacity as the computing requirements change. The Chinese Alibaba corporation also built and is expanding such an infrastructure for e-business and other multiple professions.

4. *Cloud OS*. The eyeOS is of such a kind. It reproduces the familiar desktop metaphor—with icons for files, folders, and applications—all living in a browser window. With decades of development, cloud computing has achieved great success in commercial market. Nowadays, Amazon's EC2, Google's Compute Engine (GCE)[4], and Microsoft's Azure[5] are typical examples of cloud computing.

Pervasive computing

In recent years, the integration of cyber space and its physical counterpart is becoming much closer. Smart devices with enhanced processing abilities and network-intensive environments enable a new computing paradigm — pervasive computing. Therefore, pervasive computing is characterized as the one saturated with computing and communicating capabilities, and integrated with users so that it becomes a "technology that disappears" [159].

Pervasive computing is a major evolutionary step from distributed systems and mobile computing. Therefore, it covers the traditional mobile communication techniques, micro computing device manufacturing techniques, and software methodologies. It also has its own characteristics. The evolution is mainly reflected in the following aspects [159].

- Effective use of smart spaces. The fusion of the physical space and the cyber space enables sensing and controlling of one world by the other. Software on a user's computer may behave differently, depending on where the user is currently located. Smartness may also extend to individual objects, whether located in a smart space or not.

- Invisibility. If a pervasive computing environment continuously meets user expectations and rarely presents him with surprises, it allows him to interact almost at a subconscious level.

---

[3]https://aws.amazon.com/ec2/

[4]https://cloud.google.com/compute

[5]https://azure.microsoft.com/

- Localized scalability. With the growing complexity in the smart space, the intensity of interaction between a user's personal computing space and his/her surroundings also increases. This has severe bandwidth, energy, and distraction implications for a wireless mobile user.

- Masking uneven conditioning. There will persist huge differences in the "smartness" of different environments. The variability between them should be reduced in order to improve users' satisfaction. This also implies the human-centrism philosophy in pervasive computing environments.

In [157], Saha and Mukherjee provide a pervasive computing model which consists of pervasive devices, pervasive networking, pervasive middleware, and pervasive applications. Pervasive devices include traditional input devices, wireless mobile devices, smart devices, and sensors. With the tremendous growth of pervasive devices, their communications are becoming more and more important, and this fact leads to pervasive networking. Pervasive middleware will mediate interactions with the networking kernel on the user's behalf and will keep the users immersed in the pervasive computing space [157]. Pervasive applications are the services that directly interact with users. Unlike traditional applications, they are more human-centric. This requirement will guide the middleware and networking issues to a large extent [157].

The vision of pervasive computing raises huge technological challenges. Many traditional mobile computing issues and new ones, such as scalability, heterogeneity, integration, and invisibility, are all tough to handle. To address these issues, there are many research projects in these fields. Some representative projects are Easy Living from Microsoft [161], Oxygen from MIT [155], Gaia from UIUC [153], Aura from CMU [78], and so forth.

In fact, the pervasive environment shares a lot of common characteristics with the open environment, and the research on adaptation in a pervasive environment can shed new light on that in an open environment.

Aspect oriented programming

Gregor Kiczales defines aspect oriented programming (AOP) as a new evolution in the line of technology of separation of concerns — technology that allows design and code to be structured to reflect the way developers want to think about the system [60]. AOP is based on the idea that computer systems are better programmed by separately specifying the various concerns of a system and some description of their relationships, and then relying on mechanisms in the underlying AOP environment to weave or compose them together into a coherent program. Concerns can range from high-level notions like security and quality of service to low-level notions such as caching and buffering. While the tendency in object-oriented programming is to find commonality among classes and push it up in the inheritance tree, AOP attempts to realize scattered concerns as first-class elements and eject them horizontally from the object structure [61]. As

AOP emphasizes more on the separation of concerns in software design, it supports cleaner organization of software modules and higher reusability. Some successful applications of AOP technology include JBoss application server, spring framework, and .Net framework, etc.

AOP can be basically classified into two categories, *i.e.*, static AOP and dynamic AOP, according to the weaving time and the style of concerns. In the early stages, some prototypes of AOP systems are mostly static, *e.g.*, AspectJ [107]. In static AOP, weaving the aspect is part of compiling process. For example, in AspectJ, the operation is at the byte-code level. After compiling, the result is the common java byte-code. The main advantage of this type is its simplicity to support the aspect weaving. However, the disadvantage is that it does not support the dynamic loading of new aspects. This problem leads to the introduction of dynamic AOP which supports more flexible aspect weaving. For example, Spring framework utilizes a specific proxy module to manage the loading and weaving of aspects.

The AOP based adaptation has two types. The first is to encapsulate the adaptation logic into the aspects. By identifying the adaptation concerns, AOP provides an effective means to modularize both application independent and application specific facets of adaptation. Examples can be found in [148]. The other kind of AOP based adaptation is the addition and removal of the aspect itself during execution. As aspects encapsulate specific concerns, functional or non-functional, dynamic loading and removal of aspects can cause the behavior adaptation of applications.

Discussion

We have described several novel computing paradigms and programming methodologies that are related to the software adaptation in an open environment. Technology advances and new requirements have given birth to the emergence of these novel computing paradigms and methodologies. Grid computing, cloud computing, and pervasive computing provide new computing environments. AOP provides a new programming methodology which supports higher reusability. Service oriented computing offers an insight on how to organize heterogeneous software and mask the underlying heterogeneity. The autonomic computing initiative recognizes the importance of self-management for future software and provides a basic research framework. In fact, because of the similarities in the background of these paradigms, the boundary between them is blurring nowadays. Many techniques overlap with each other.

For these computing paradigms, adaptation issues are also addressed. For example, in the grid environment, because the computer node may enter or exit the grid randomly, self-adaptation algorithm is necessary to cope with the uncertainty. Despite the similarities with the grid, cloud, and pervasive computing, an open environment has its own features. It requires systematic techniques and theory [187, 120] which include requirement definition, software architecture design, analysis, adaptation techniques, and maintenance, etc. This book mainly address the adaptation concerns raised by the open environment. Considering the broad area adaptation covers, we particularly focus on context modeling, adaptation enabling techniques, adaptive component migration, connector-based service discovery and adaptation, and modeling/analysis aspects.

1.2   What is Self-Adaptation?

Self-adaptation is a compound. So first let us examine *adaptation*. In the *Longman Dictionary of Contemporary English*, adaptation has two meanings:

1. A film or television program that is based on a book or play;

2. The process of changing something to make it suitable for a new situation.

Evidently, our usage of *adaptation* fits the second explanation. Generally, it denotes a process that the entity can adapt itself according to the environment. As aforementioned, adaptation involves multiple software engineering disciplines. Many researchers have given their own definitions from different perspectives. For instance,

■ A program can change its behaviors automatically according to the context [118].

■ Being able to make last moment changes [3].

■ Any automated and concerted set of actions aimed at modifying at runtime, the structure, behavior, and/or performance of a target software system, typically in response to the occurrence and recognition of some (adverse) conditions [177].

From the above definitions, we can observe that, except for the second one which is defined from the software process dimension, the other two share similar elements, *i.e.*, **contextual information** and corresponding **runtime reactions**.

Self-adaptation emphasizes more on the autonomy. DARPA Broad Agency (BAA-98-12) provides a definition of self-adaptive software as follows:

Self-adaptive software evaluates its own behavior and changes behavior when the evaluation indicates that it is not accomplishing what the software is intended to do, or when better functionality or performance is possible.

This definition is mainly given from an artificial intelligence perspective. In our opinion, self-adaptation includes the following key factors.

■ First, it is an interaction between software and the environment. The software probes the changes from the environment and reacts accordingly.

- Second, it is an intelligent behavior. Software can choose the embedded reaction strategies or load them dynamically based on the specific context.

- Third, it has a realtime feature. Adaptation should be performed without causing system's shutdown.

- Fourth, it is a purposiveness process. It reflects the design aim, *e.g.*, to provide continual availability or to improve users' satisfaction.

In light of these considerations, we present our definition of self-adaptation as follows.

> *Self-adaptation is a process through which software entity dynamically changes its behavior according to the context information in order to ensure or improve users' satisfaction.*

There is a subtle difference between *adaptation* and another buzzword, *i.e.*, *online/dynamic/runtime evolution*. In [182], Wang *et al.* define online evolution as a specific kind of evolution that updates running programs without interrupting their execution. From this definition we can observe that online evolution mainly focuses on the techniques that can update a system dynamically which is the third factor listed above. But self-adaptation also covers other dimensions, such as, interaction with the context and the decision-making process.

Software adaptation can be inspected from different perspectives. Therefore, there are several classification criteria. In [151], Rohr *et al.* give a comprehensive classification schema. This tree-like schema examines software adaptation from five dimensions, *i.e.*, *origin*, *activation*, *system layer*, *operation*, and *controller distribution*.

- *Origin* is the location of the state change that triggers an adaptation cycle. The changes may take place in the external environment or in the systems themselves.

- *Activation* denotes the response types. There are three types, *i.e.*, reactive, predictive, and proactive. Reactive adaptation means that the system adapts only after that performance has degraded and is below a certain threshold. Predictive adaptation means that before the real drops of performance, some prediction algorithms can foresee the potential threats and respond beforehand. Proactive adaptation is mainly applied to improving system performance when it is still normal with no signs of potential drops. This type of activation is closely related to self-optimization.

- As computing systems are hierarchical, *system layer* denotes the corresponding adaptation layers. Generally, there is a hardware level, an operation system level, a middleware level, and an application level. Self-adaptation techniques can be applied to any of these layers.

■ *Operation* classification dimension uses an architecture description language viewpoint on the adaptation operation. Multiple component instances can belong to the same component type, and the implementation of a component instance can change without affecting the associated component type. Therefore, six types of adaptation operations are distinguished, *i.e.*, data adaptation, intra-component behavior adaptation, component resource mapping adaptation, inter-component protocol adaptation, instance-level adaptation, and type-level adaptation.

■ *Controller distribution* denotes the localization of self-adaptation logic. Generally, there are three types, *i.e.*, centralized, decentralized, and hybrid. The controller distribution style is centralized if the adaptation is controlled and executed from a central point in the system, while a decentralized system has no such central control module. Hybrid approaches combine the advantages of centralized and decentralized architectures by applying parts of the activities locally and other parts globally [151].

Rohr's classification schema is based upon the different stages of adaptation process. Similarly, Subramanian gives a spectrum of adaptability phenomena [170] based upon the more fine-grained self-* facets, *e.g.*, self-reparability, self-tuning, etc. Self-reparability is the ability of software systems to adapt at runtime in order to accommodate varying resources, system errors, and changing requirements. Self-configuring systems can configure themselves in the field rather than in the factory. Self-tuning can tune and reconfigure themselves. Autonomic systems are capable of running themselves, adjusting to varying circumstances, and preparing their resources to handle most efficiently the workloads put upon them. Extensibility is the case with which a system or component can be modified to increase its storage or functional capability. Tailorable systems can be adapted by the users to their particular situations and needs. Changeability is the ability of a software system to sustain an on-going flow of changes. Modifiability is the ability to make modifications to the software system. Evolvability enables easy evolution of software systems through enhancement to meet current needs. Flexibility has been defined as the ease with which a system or component can be modified for use in applications or environments other than those for which it was specifically designed [170].

It should be emphaized that this spectrum is not static. New or similar self-* terms are under development, such as self-management [28], self-optimization [4], and self-healing [81]. Self-management systems denote those systems that not only implement the change internally but also initiate, select, and assess the change itself without the assistance of an external user [28]. Self-optimization refers to the automatic mechanism for recognizing significant changes and re-optimizing the system as a result of such changes [4]. Self-healing is defined as the property that enables a system to perceive that it is not operating correctly. Without (or with) human intervention, it can make the necessary adjustments to restore itself to normalcy [81]. From the above introduction, we can observe that, in fact, there are no sharp distinctions between self-* concepts. Instead, there is only a subtle difference in the aspect of adaptation emphases.

1.3   What is Context?

Context is an intensively studied subject in recent years [13]. In the *Longman Dictionary of Contemporary English*, context has two meanings:

1. The situation, events, or information that are related to something and that help you to understand it;

2. The words that come just before and after a word or sentence that help you understand its meaning.

As the second explanation mainly deals with grammar, our usage corresponds to the first one. Many researchers have given their definitions on this subject. Brown *et al.* define context as a collection of locations, identities of the people around the user, the time of the day, the season, the temperature, etc. [31]. Ryan *et al.* define context as a synonym of environment, which includes the information such as the location, the time, the temperature, and the user identity [156].

In [55], Dey gives a widely accepted definition: *Context is any information that can be used to characterize the situation of an entity. An entity is a person, place, or object that is considered relevant to the interaction between a user and an application, including the user and application themselves.*

This definition is proposed from the pervasive computing perspective [159]. However the definition is too broad. We believe that for self-adaptive software, the context should be narrowed down to those factors that have potential influences on the software's adaptive behaviors. Thus we give our definition of context as follows:

> *Context is the relevant information that can potentially cause or constrain software's adaptation.*

There is a need to differentiate this term with *environment*. These two concepts obviously have overlaps. On the one hand, environment definitely has potential influences on software's adaptation, but not every piece of information from the environment makes sense in this process. On the other hand, environment information is not the only factor affecting adaptation, other factors also matter. As a concrete example, a web application responds slowly and requires adaptation to add another backup server. If the architectural style specifies that at most N backups can be added, the current architecture information and the style constraints are both counted into context.

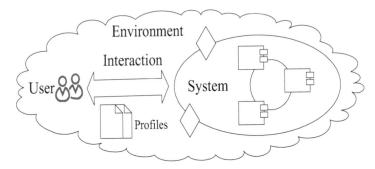

Figure 1.3: Interactions among user, environment, and system

1.4 Challenges of Adaptation in an Open Environment

In an open environment, the research on self-adaptation software faces many new challenges due to the increasing complexity of the context. As the traditional methodology and theory for software adaptation is developed in a relatively closed, static, and controllable environment, some of them are not applicable. For instance, traditionally, software architecture is regarded to be static throughout the life cycle of the software system. However, this assumption does not hold under the new circumstances. Instead, in many cases, flexible adaptation of software architecture is needed to meet the requirements of the context changes [121].

As the new paradigm emerges and the environment becomes more and more open, the context is much more complex than ever before. An understanding of the interaction patterns between software systems and their environment is essential to construct self-adaptive software in such environments. Figure 1.3 illustrates an abstract view of the interaction. In this process, factors from the user, the interaction manner, and the environment all possibly affect the running system. These factors constitute the essence of context that causes software adaptation.

Having recognized the importance of context, the next issue is how to model it and provide a basis for application adaptation. An adequate and unified context model plays an important role in self-adaptive systems. Firstly, it can enhance the representation ability and facilitate the following reasoning and reacting process; secondly, a unified model can help avoid communication problems between different applications, which are usually caused by a lack of common understanding.

After getting the necessary information from the context, the system needs to adapt itself accordingly. This requires an adaptation enabling mechanism that can update the system dynamically. The process requires no interruption of the application. As in an open environment, systems are commonly composed by a collection of loosely coupled components, the adaptation mechanism is usually reflected in the coordination entity.

In [110, 137], Kramer *et al.* and Oreizy *et al.* stated the essential role that software architecture plays in self-adaptation. Respectively, they both proposed reference frameworks. However, the proposed frameworks stay at a very high level and share a lot in common. Nevertheless, they shed insight into the research on software adaptation. For example, they both use a control loop-like mechanism, and separate the adaptation logic from the business logic. However, the problem is that both of them are very abstract. In concrete scenarios, it is difficult to apply them directly. Moreover, both of them lack an explicit context model.

1.4.1 Characteristics of the open environment

Compared with traditional software adaptation research, in an open environment, adaptation faces new challenges. According to the interaction paradigm described in Figure 1.3, we will examine the characteristics and requirements from three perspectives.

1. The open environment has the following salient features.

    (a) Dynamics. This feature is reflected in the randomness of the computing node availability, existence, or the network conditions. The adaptation should handle the cases that the involved application components operate on these dynamic nodes and networks.

    (b) Openness. The new node may join or exit the existing network without notification. Thus, the size of the network may increase or decrease randomly. The adaptation should have the ability to incorporate or remove corresponding components.

    (c) Diversity of platforms and devices. With the development of technology, computers with high processing abilities are diversified into common consumer electronics which have different kinds of properties. Applications should adapt their behavior when running on different platforms. For example, some graphical display functions may have to be adjusted when the application migrates from a PC to a hand-held device due to their different display sizes.

    (d) Heterogeneity of protocols. The sub-network may have various different protocols. Standards may help to eliminate some heterogeneity, but the situation is there are usually several co-existing standards which still leads to the problem of heterogeneity. It requires that the coordination mechanism is flexibly adapted to support multi-mode protocol interactions.

2. The structure patterns of current software in such environments are different from those previously constructed.

(a) Distribution. The components are usually scattered on the net, provided as a stand-alone service or an elemental service that can be integrated by third parties. These distributed components do not necessarily belong to the same organization. They only provide some public interfaces for communication. Thus, the integrators have limited control over these distributed components.

(b) Coarse-grained components. The component size is more coarse-grained and self-contained. They can be invoked directly through some standard protocols or composed by third parties. Service oriented computing is emerging and offers a very promising business solution [100]. The adaptation of such systems is usually in the form of composition reconfiguration.

(c) Various interaction modes. As the integrator has limited control over the behavior of the components, the coordination mechanism becomes more important. Common interaction methods, such as RPC, and tuple space, implicitly know about the potential interaction entities. In multi-mode interaction context, it requires a more flexible coordination mechanism to adapt accordingly.

(d) Trust management issues. Usually, these services cannot guarantee the service quality and robustness. Even worse, some fraud services exist. To adapt with trust concern, this requires the deep analysis of the interaction confidence and the entity's historical reputation.

3. With the growing intensity of computation and networking, computing is more human-oriented [155]. Factors from the interactions between users and computers also pose new challenges to software adaptability.

(a) Interaction profile. This usually includes personal preferences, operation habits, etc. Applications need to adapt according to the users' special needs in order to improve their satisfaction.

(b) Users' locations. Location is a concept closely related with mobility. This is an active research subject of context-aware computing [13]. To ensure continual availability, some applications need to migrate along with the users, such as editors or media players. This requires application migration. Complete migration costs are high. Therefore a light-weight, adaptive component-level migration is much more desirable.

1.4.2   Adaptation requirements

As mentioned before, the open environment poses new challenges to the self-adaptation of inhabitant software applications. We use a concrete example to identify some key adaptation requirements. Consider the following scenario: a user is interacting with a data report application at his office. The application is composed of several components. The schedule alerts that, in several minutes, he is supposed to give a presentation about his work in a conference venue. When time is up, he has to leave the office, but he wants to continue working on the report on the way to the conference room. The sensor detects his movement and transfers the corresponding components of the application to his smart device. In this way, the user can continue editing. After arriving at the conference room, the sensor detects his arrival and restores the application component to the desktop computer in the new environment and establishes the connection with the original components. As the conference room is in a different network from the office, it is highly desirable to add a new security component on the fly. In the conference room, the user might need the printing service. However, the service protocol is not necessary the same as that in the old environment. For example, the printing service protocol in the office network is Jini and in the meeting room it might use the UPnP protocol.

The scenario reflects many characteristics of openness. For example, the fact that the application runs across networks is a kind of boundary openness. For the conference network, the user is new. He is eligible to use the network and hardware devices. This is a kind of usage openness. Different service protocols and devices co-exist and co-operate in the different networks. This is a type of protocol openness. In this scenario of open environment, we can identify some key self-adaptation problems.

1. In an open environment, the context is more complex and variant than before, how to understand and model the context information in order to facilitate the following adaptation process;

2. Because of the boundary openness and users' mobility, how to adapt the application (migrate with the user) accordingly in order to provide continual services;

3. In an open environment, multiple service discovery and interaction protocols might co-exist, how to adaptively handle this heterogeneity and support the multi-mode interactions flexibly;

4. The richness of the context information makes multiple adaptation rules concurrently available possible, how to check the relationship between the adaptation rules in order to ensure that the rules themselves are not in conflict with each other;

5. In the adaptation cycle, how to assure the dynamic evolution process is consistent with the user's expectation and how to model and analyze the decision-making process.

The above list gives some key self-adaptation problems, not only because these problems are essential in this scenario, but also because they are common and representative in an open environment. One one hand, each problem is standalone and attracts many research efforts; on the other hand, these problems have underlying connections. To provide continual service during users' mobility, application migration is desirable. During the migration process, it has to cope with the possible underlying heterogeneity. As these actions happen in an open environment, the context is much more complex. To ensure an effective adaptation, a careful study on the context model, adaption enactment mechanisms, the adaptation rule management, and related formal models is highly necessary.

1.5   Structure of the Book

Starting from the concrete scenario given above, in this book, we inspect the identified problems and attempt to develop more systematic solutions from a perspective of software architecture, including context modeling, adaptive service discovery and multi-mode interaction, adaptive component level migration, adaptation rule conflict detection, dynamic evolution verification, and Markov Decision Process (MDP) based decision-making process. The book is organized into three parts.

- The first part, which consists of four chapters, provides some basics and an overview of architecture based self-adaptation techniques. Chapter 2 presents the conceptual framework for self-adaptation in an open environment. Chapter 3 introduces the ontology-based context modeling techniques, and Chapter 4 presents a middleware platform to support the framework.

- The second part deals with specific issues of adaptive application migration and service interaction. It consists of two chapters, Chapter 5 and Chapter 6. Chapter 5 explains the solution for adaptive component-level application migration with users' mobility in the open environment, while Chapter 6 presents the connector-based adaptation techniques for service discovery and multi-mode interaction in heterogeneous networks.

- The third part of the book covers the issues of formal modeling and analysis aspects of self-adaptation process. It consists of three chapters. Chapter 7 introduces the graph transformation-based adaptation modeling. On top of this, multiple architectural adaptation rules can be checked formally and the conflict/dependency relations can be detected. Chapter 8 introduces the techniques of modeling and verification of the enactment part of self-adaptation, *i.e.*, dynamic evolution, so as to assure the adaptation is consistent with the specification. The formal modeling and analysis of decision-making process of self-adaptation is given in Chapter 9.

In the sequel, we would like to highlight some results which are achieved by the authors in the past 10 years via collaborations with other researchers. Some results were published in the form of research articles, but were somehow dispersed in the literature. Here, for the first time they are collected systematically in a single book.

1. An ontology-based context model is proposed. The necessity of explicit context modeling in an open environment is clarified. By introducing ontology into the architecture level and combining with our previous work on the dynamic software architecture, the semantic gap between external context models and the internal architectural specification is bridged. Therefore, the selection of adaptation strategies can refer to the architectural knowledge. Moreover, a basic conceptual framework for software adaptation is discussed. The framework can help to understand the related concepts, and techniques of adaptation in an open environment.

2. An adaptive component migration mechanism is proposed. With the loosely coupled application model and ontology-based context reasoning technique, the mechanism can support adaptive component-level migration. We also leverage attributed graph grammar to specify the changes of the architectural deployment view during the migration process. In this way, some properties, such as the deployment constraints, can be checked formally. Compared with previous work, the adaptive approach can reduce the network load and the response time.

3. A connector-based adaptation approach for heterogeneous service discovery and multi-mode interaction is proposed. In response to the heterogeneity characteristic of an open environment, the approach which is combined with the context-aware techniques can support service discovery and interaction in heterogeneous sub-networks. Preliminary experiments demonstrate the effectiveness of the approach.

4. A critical pair analysis based approach is proposed to detect the conflict and dependency relations of concurrent architectural adaptations. The architectural style is formally specified by the attributed graph grammar, and the reconfiguration adaptation is expressed by graph rewriting. Critical pair analysis can be used to statically check the relations between the rules. Four categories can be classified, *i.e.*, parallel, dependent, asymmetric conflicting, and symmetric conflicting. With the graph grammar directed architecture development environment, the mechanism can help ensure that the software architecture conforms to the constraints of a specific architectural style in the process of composition and adaptation.

5. An integrated development environment, MAC-ng, is presented. MAC-ng is a middleware platform that supports architecture-based self-adaptive

software development in the open environment. It supports the ontology-based context modeling, reasoning, and the following architectural adaptation. The conflict detection module and the grammar directed editor are integrated with the platform.

6. A case study on the water management application is discussed. Water management usually involves a wide scope, diverse platforms, and complex context information. The thesis attempts to apply some of the proposed approaches to this domain. An experimental prototype is implemented. The experiments demonstrate the feasibility of these proposed approaches.

7. A behavioral modeling and verification approach for the dynamic evolution process is presented. Dynamic evolution is the underlying enabling mechanism of self-adaptation. To keep the evolution process consistent with the specification is a prerequisite of self-adaptation. Based on architecture characteristics in an open environment, we present a hierarchical timed automata based approach to model and verify the dynamic evolution process during adaptation.

8. An iterative decision-making scheme for self-adaptation is presented. The scheme infers both point and interval estimates for the undetermined transition probabilities in a Markov Decision Process (MDP) based on sampled data, and iteratively computes a confidently optimal scheduler from a given finite subset of schedulers. The most important feature of the scheme is the flexibility for adjusting the criterion of confident optimality and the sample size within the iteration.

Chapter 2

Adaptation Framework

CONTENTS

2.1 Introduction and Background

There is a wide range of techniques that can enable dynamic adaptation. In early times, the adaptation logic was mainly embedded into the application code and was provided as a domain specific solution to adaptation requirements. Different concerns are entangled, and, thus, the system supports poor reusability and flexibility. In addition, since the control logic is hard-wired into the application, when events which are not predefined occur, the system has to be redesigned. Regarding these shortcomings, more and more work separates the adaptation concern from the business concern. Generally speaking, for this paradigm, the adaptation control part probes the context information and acts upon the system, which forms a closed loop.

With regard to the important role that the architecture plays in software engineering [175], a lot of work attempts to introduce dynamics into the software architecture. Many Architecture Description Languages (ADLs) have been proposed to describe the software architecture, and some automated tools have been designed to help generate the skeleton of the system [28]. The changes are first acted upon the models, propagated, and then finally acted upon the destination system by a newly generated skeleton. The system is thus updated. As the architecture is a global abstraction of the system, it can reflect the functional or nonfunctional requirements of the design. The architectural level adaptation is usually on a large scale compared to other types, for example, on the algorithm level or the parameter level. In open environments where more and more applications are composed of a set of coarse grained, self-contained components, architecture-based adaptation offers a promising approach. Architectural reconfiguration can be used to model the dynamic addition, removal, or update of the constituted elements [136].

Kramer and Magee present a three-layered architecture model for self-management [110], *i.e.*, *goal management, change management,* and *component control*. The status report from the component control layer will be reported upwards to the goal management layer, then the change plans derived from the goal management layer lead to change actions in the change management layer, and in turn the change actions will affect component connection in the component control layer. Similar to this, Mei *et al.* propose a software architecture-centric approach for software adaptation [127]. The proposed models and approaches focus on the techniques that can guide dynamic changes to the system in a disciplined way. We believe that in an open environment, architectural models can offer the required level of abstraction and generality, but the complexity of the context should have been given more considerations, and the mechanism that bridges context models and architectural models should also be studied.

Software adaptation is closely related to the architecture reconfiguration. Because of the generality, levels of abstraction, and potential for scalability provided by the software architecture, many researchers approach the adaptation issues from this perspective [110]. Generally, these approaches include architectural specifications and an enabling mechanism for the specification which provides a basis for self-adaptation. It also includes the definitions of architectural styles and multi-views of the architecture, etc. In this section, we will discuss these related concepts and concerns in architecture-based software adaptation.

2.1.1 Architecture description languages

Architecture description languages (ADLs) provide a means to model and analyze software architectures in order to support the architecture-based software development, and improve software quality and correctness. Existing ADLs can be classified into two categories, *i.e.*, general-purpose and domain specific ADLs. These ADLs provide support for structure abstraction and analysis, but they are different from common programming languages. They focus on the global behaviors instead of partial ones. Generally, all ADLs share some common characteristics, such as a specification language with formal or informal graphical notations, distributed system modeling support, treating data flow and controlling flow as interconnection mechanisms, and layer support, etc. In general, almost all ADLs feature the following three elements:

Component Component denotes the unit for computation and data storage. Components are characterized by their interfaces, types, behaviors, and constraints, etc.

Connector Connector describes the interaction between the components. Most ADLs regard the connector as an explicit and first-class element in software architecture. Similar to components, connectors are also characterized by their interfaces, types, behaviors, and constraints, etc.

Configuration Configuration specifies how to connect the components and connectors. The topology describes the global structure. Configuration description is relatively independent from the concrete components and connectors. Besides, most of the ADLs provide layered descriptions of the global structure. Multiple components or connectors can be encapsulated into a compound component or connector.

Certain software architectures share similar configurations and constraints. This fact leads to the concept of architectural style. Architectural style is a specific family of software architectures. They are characterized by the structural patterns of the organization of the components and connectors. Generally, an architectural style determines the vocabulary of components and connectors that can be used in instances of that style, together with a set of constraints on how they can be combined [162]. In this sense, software architectural style provides a basic guideline for the adaptation constraints.

2.1.2 Software architectural views

Software architecture can be viewed and analyzed from different perspectives. Kruchten proposed a 4+1 model to describe software architecture using five concurrent views [111]. The first four views are *logical view, process view, physical view,* and *development view.* The last view is *use cases or scenarios.* Each view addresses a specific set of concerns of interest of different stake-holders in the system.

- Logical view primarily supports the functional requirements. The system is decomposed into a set of key abstractions. These abstractions are objects or classes.

- Process view focuses on the non-functional requirements, such as, performance and system availability. It addresses concurrency and distribution, system integrity, and fault-tolerance.

- Physical view also addresses the non-functional requirements, but it works on different layers from the process view. It maps the software to the computer nodes in the network.

- Development view describes the organization of the actual software modules in the software-development environment.

- As a complement to the above four views, scenario is an abstraction of the users' requirements. In the design phase of software architecture, scenarios can help to identify architectural elements and be used to validate the system afterwards.

2.1.3 Software architecture dynamics

The *dynamic* software architecture models a system that reacts to certain events at runtime by supporting reconfiguration of the system's architecture. The software architecture dynamics have two levels. The first level is the modeling of the adaptation, and the second level is the enactment mechanism for the adaptation. A majority of existing ADLs view configurations statically, and some others support only constrained dynamic manipulation of architectures. However, more research on the modeling of dynamic software architecture has been conducted recently.

Process algebras Magee and Kramer proposed a language called Darwin to describe distributed system structure. The operational semantics for the dynamic features is presented in the π-calculus [122]. Darwin uses two techniques to capture the dynamic structure. The first is lazy instantiation. The second technique is direct dynamic instantiation which allows arbitrary structures but permits the context or environment of the components to be precisely captured in the structural description. Recently, the labeled transition system (LTS) is introduced into the language to support the description of the system's behavior. Similar to Darwin, Wright which is developed at CMU, is based on the CSP. The dynamics are realized by an external configurator. Furthermore, other calculus systems, such as join calculus, and HO-π calculus, provide support for adaptation.

Event system The representative language of this kind is Rapide. Rapide is an event-based concurrent object-oriented language specifically designed for prototyping architectures of distributed systems. Another example is C2 style ADL. C2 style architecture communicates by selective multi-casting. It introduces the notion of event translation: a transformation of the requests issued by a component into the specific form understood by the recipient of the request, as well as, the transformation of notifications received by a component into a form it understands. C2 style regards a system as a hierarchy of concurrent components bound together by connectors. Components within the hierarchy can only be aware of the components "above" it but is completely unaware of components residing at the same level or "beneath" it, and, at runtime, C2 can add, delete, or rearrange components [136].

Chemical abstract machine CHAM was firstly proposed by Berry and Boudol based on the chemical metaphor [22]. The state of a system is like a chemical solution in which floating molecules can interact with each other according to reaction rules and a magical mechanism stirs the solution, allowing for possible contacts between molecules. In the early stages of CHAM, it did not support the modeling of dynamic adaptations. Later, Wermelinger [184] tried to combine the CHAM and graph grammar to provide a comprehensive framework for the modeling of software architecture adaptation.

Graph grammar Graph notations have been widely applied in software architecture modeling [18, 28, 115] due to their unique characteristics. Firstly, graph representation has the advantage of an intuitive visual correspondence to the common practice of box-line notation of software architecture; secondly, the widely recognized importance of a connector as a first-class entity can be well captured by the notion of edge attributes; thirdly, by *production*, the graph rewrite techniques can be employed to describe the evolution and to check some interested properties, such as, confluence and conflicts formally.

Other approaches Stuurman *et al.* proposed two mechanisms for dynamic adaptation in the distributed control system. The software architecture used is a configuration of distributed processes, communicating according to the subscription model. One of the mechanisms is associated with the decoupling of processes as a result of the subscription-based communication model. The other mechanism is based on the late-binding properties of Java [168]. Mei *et al.* proposed an architecture-centric adaptation approach. The monitor, analysis, planning, and enactment all center around the software architecture. The approach incorporates a set of techniques, such as, reflection and reasoning, to provide a comprehensive methodology for self-adaptation in open environment [127]. The approach is promising, but it is still in the early stages.

2.1.4 Discussions

Software architecture provides an overall view of the structure, behaviors, and properties of the applications [174]. Since the design is the central focus of software engineering, software architecture is becoming the most crucial artifact of this process. Regarding its importance, numerous ADLs have been proposed to describe and analyze the architecture properties. Architectural styles have been identified to model a family of software architecture instances which share common characteristics. Most ADLs in the early stages only addressed static modeling issues. As the need for the dynamic adaptation is growing, research emphases have gradually been shifted to the dynamic aspects.

Traditional architectural view models, such as Kruchenten's 4+1 view model, do not address much about the dynamic aspects of the architecture. Due to this insufficiency, Oquendo *et al.* [135] proposed a behavior view to specify the functions and behaviors of software systems. Behaviors are further classified into two types, *i.e.*, computation behavior and adaptation behavior. The computation behavior addresses the business logic, while the adaptation behavior addresses the dynamic aspects. This separation of views facilitates the specification of dynamic behaviors of software architecture.

2.2 Conceptual Adaptation Framework

2.2.1 Architectural perspective

Software architecture offers a coarse-grained view of the system at the level of interacting components and connectors. Architecture-based adaptation is reduced to the adaptation of the constituent components and connectors and their reconfigurations. Compared with other *ad hoc* solutions, architecture-based adaptation has a better reusability and tractability. Since components usually contain functional codes while connectors contain coordination counterparts, this separation of concerns promises a vision of less difficult consistency control during reconfiguration.

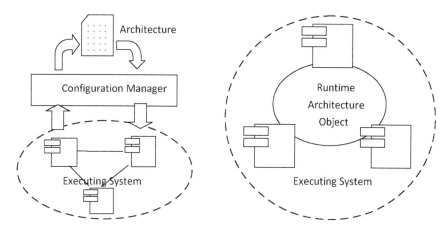

Figure 2.1: Contrast with external approach

Realizing the generality and the appropriate level of abstraction of the software architecture, a proliferation of work has emerged to study its support for adaptation [110]. A considerable amount of such work approach the problem from an external description perspective by using some architecture description languages (ADLs) to specify the architecture [126]. Nevertheless, most of them stay at the level of specification and a few are accompanied by a strategy for implementing the described architecture which in turn is hard to ensure that the implemented system conforms to the architecture [125]. As in traditional approaches, the adaptation is first conducted at the ADL level, generating a new skeleton of the system, and then the system is reconfigured and restarted. It requires a specific architecture configuration manager to connect the specification and the running applications. The configuration manager needs to distill the architecture information dynamically and conduct adaptation to the application. This difficulty limits its application to a greater extent. Besides, as the architecture reflects the set of design decisions about the system, this information is not retained in the running applications. To address these problems, we proposed an internal architectural object based approach to explicitly maintain the architectural information [121]. Figure 2.1 illustrates the difference between our approach (right box) and the traditional specification-based approach (left box).

In essence, this approach is from the view of coordination. Architectural object has the topology information of the system and intercepts the method invocations. Concisely speaking, the context changes are transmitted to the architectural object and cause its update. As the architectural object contains the interaction logic, the invocations between the components are reinterpreted. In this way, the system is dynamically evolved. The detail of this approach is explained in the next section.

As software architecture contains a set of components and connectors, its adaptation scale includes system level adaptation, component level adaption, and connector level adaptation. By component level adaptation, we mean that the component changes its behaviors or other properties dynamically, such as, the location. Since we focus on coarse-grained adaptation, the adaptability inside the component is left to the specific self-adaptive algorithm design and is outside our concern. Thus the component level adaptation denotes the dynamic changes of the component properties, such as, the locations and their compositions. By connector level adaptation, we mean that the connectors can adaptively adjust their composition or distribution to meet the coordination needs posed by the environment. System level adaptation means the general reconfiguration of the involved components and connectors, and the dynamic addition and removal of the constituent components and connectors. As the component level and connector level adaptations can be generalized into the system level reconfiguration, these three levels are connected closely with each other, and we do not distinguish them deliberately.

In [110], Kramer and Magee summarize that architecture-based adaptation offers several potential benefits.

1. Generality - the underlying concepts and principles should be applicable to a wide range of application domains.

2. Level of abstraction - software architecture can provide an appropriate level of abstraction to describe dynamic change in a system, such as, the use of components, bindings, and composition, rather than at the algorithmic level.

3. Potential for scalability - architectures generally support both hierarchical composition and hiding techniques which are useful for varying the level of description and the ability to build systems.

4. Builds on existing work - there is a wealth of architecture description languages and notations which include support for dynamic architectures and for formal architecture-based analysis and reasoning.

5. Potential for an integrated approach - many ADLs and approaches support software configuration, deployment, and reconfiguration.

The above are the general benefits of the architecture-based approach. In an open environment, we also have the following concerns.

1. The components are usually distributed in the network and are not controlled by the integrators. These coarse-grained entities cannot guarantee their service availability. Besides, the system can dynamically search other components and integrate them into the system.

2. It is not uncommon that some sub-networks are heterogeneous, and they

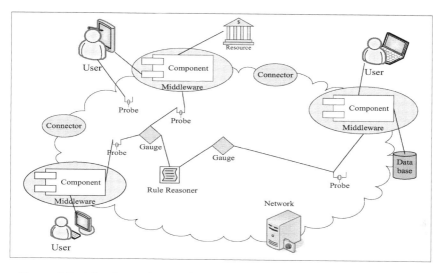

Figure 2.2: Adaptive software elements illustration in an open environment

use different protocols to interact. These different kinds of service discovery and interaction parts are usually encapsulated as coordination entities (connectors). To support inter-communications, protocol-level adaptation can be mapped to the connector-level adaptation. It supports flexible interactions.

3. In an open environment, there is a need to provide a non-stop follow-me kind of application migration ability for the user's mobility. Due to the feature of the resource intensiveness in open environments, it is effective to provide adaptive component-level migration. This can be mapped to the component-level adaptations.

2.2.2 Framework overview

There are various components and resources scattered in the network providing services. Different devices access these services in the open environment. Weaving them into an adaptive system requires a flexible coordination mechanism, context probes, gauges, and adaptation rules, which are illustrated in Figure 2.2.

1. Context. Context is the key factor that drives self-adaptation. It includes the user profiles, device properties, software architectural knowledge, and operating environment properties. Context is one of the central concepts in the framework, as the environment, the users, and the applications themselves are part of the context. For example, when the user's location changes, a music player component follows the user to the new environment, and the sensors report that the context is noisy. Besides the normal

requirement of continuing the music playing, it is possible that the user wants the volume to be turned up automatically according to the noise degree of the new location. In this case, the environmental condition, the network location, the user's profile, and the application architecture are all parts of the context, and they provide important information for adaptation.

2. Component. Components are modules that usually contain business logics. There are several kinds of components in a typical open environment. The first kind is the traditionally loosely coupled components, developed and maintained by the application developer and distributed over the network. The second kind is more independent networked components. They are developed by some third parties and can be integrated by others. Usually these components have self-describing specifications, and there is a central registry center for publishing and retrieving these components.

3. Connector. The coordination logic is embedded in the form of connectors. Connectors play an important role in software adaptation in open environment, especially for those independent, autonomous component based systems. As the components may disappear suddenly because of the dynamics or the discovery of better components, the connectors play a crucial part in handling such cases to ensure that the system can adapt without violating the constraints.

4. Probe and Gauge. Physical sensors or software sensors can get the data. We call these sensors probes. For example, some infra-red sensors can detect the users' location. Some modules can dynamically learn users' operation habits or just parse the static profile specifications. Gauges [77] are modules that consume and interpret lower-level probe measurements in terms of higher-level model properties. Thus, gauges are used to collect information and calculate observations that are relevant to the adaptations. For example, the distance information from the sensor can be transformed into the concrete locations, such as office, by the gauge.

5. Rule Management. It mainly includes rule reasoning and conflict detection. Rule reasoner consumes the information from the gauges and conducts the reasoning process to deduce some actions. This is a knowledge-intensive process, as it is involved with the environmental knowledge and architectural knowledge. Users define rules, and these rules become part of the reasoning engine's input. In ambient intelligence, there are several reasoning approaches, such as, the logic based approach and Bayesian network based approach [25]. The decision from the reasoning should be made upon the working systems. There should be some utility support to schedule the rules and ensure the consistency during adaptation.

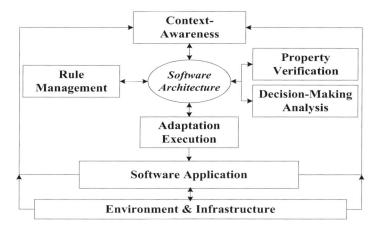

Figure 2.3: Conceptual adaptation framework

6. Assurance Mechanism. The adaptation must proceed in a disciplined way. It is necessary to leverage formal techniques to assure this process. For example, uncertainty widely exists in the open environment, it is important to infer an optimal adaptation strategy given limited information. Aside from that, the system is supposed to evolve to meet users' requirements. During the dynamic evolution phase of adaptation, it is essential that the architecture level reconfiguration strategies are consistent with the specifications. In these aspects, formal models can provide rigorous support.

The above factors are essential parts to construct self-adaptive software in an open environment. However, other basic services, infrastructure, and middleware support are still needed, such as, service publish/subscribe, security, and networking utilities, to keep applications running on top of that.

Based on the aforementioned analysis, we propose a conceptual framework for self-adaptation in an open environment. The framework follows a control-loop style and is illustrated in Figure 2.3. The context awareness module continuously monitors the states of the application as well as the environment via probes and gauges. The rule management module is responsible to manage the rules, for example to resolve possible conflicts or inconsistencies. The property verification module is to ensure the correctness of adaptation in a formal way, for example, model checking or theorem proving. The decision-making analysis module provides support for the optimal adaptation strategy inference in the face of uncertainty from the open environment. The central part of the framework is the software architecture model. It interacts with most of other modules in the framework and also directs the dynamic evolution (adaptation) process of the underlying software.

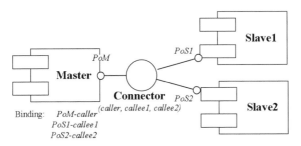

Figure 2.4: A simplistic master/slave style design diagram

2.3 Runtime Software Architecture

Context information provides the prerequisite for the adaptation. To close the loop, we still need an enabling mechanism to causally connect the environment and requirement changes with the system's behavior response. In our approach, the interplay between them is fulfilled by the notion of *polymorphic software architecture*, which consists of two essential elements, *i.e.*, software architecture class model and reflective interaction.

2.3.1 Software architecture class model

Software architecture can be regarded as "*the set of principal design decisions about the system.*" As argued previously, we believe that these decisions should be preserved and maintained explicitly in the life cycle of the system instead of just in the design phase.

Figure 2.4 further illustrates the gap between design and implementation. In a simplistic master/slave architectural style application, the *Master* component dispatches tasks to two other components *Slave1, Slave2* via port *PoM* (Port of Master). The slave components provide services via port *PoS1* (Port of Slave1) and *PoS2* (Port of Slave2). In this setting, the actual role of the *Master* is the *caller*, while that of the *Slave1* and *Slave2* is *callee*, *i.e.*, *callee1* and *callee2*. Thus, the actual design intention is to bind the *PoM* of *Master* to the role *caller* and the *PoS1, PoS2* of the slaves to the role *callee1* and *callee2*, respectively. Moreover, the design conveys two implicit decisions. First, the inter-component invocation should be coordinated through a first-class entity *Connector*; second, the binding relationship allows for an evolution space of future possible updates. However, when down to implementation, the coordination is dispersed into the tight component reference and the associated method invocation, which obviously hampers the system's maintainability. For example, if another component with more efficient implementation substitutes *Slave1*, all the references to *Slave1* in *Master* have to be changed instead of the simple update of binding information of the *Connector* as intended in the design.

The above consideration motivates us to keep an explicit architectural model up-to-date during system runtime. The model is instantiated by the *architecture class* which reifies the concept of an architectural style. All architecture classes inherit from a common parent class RuntimeArchitecture directly or indirectly. RuntimeArchitecture provides the basic functions for the development of a specific architecture class, including: 1) architectural topology, which is a canonical programming-level representation of the software architecture specification; 2) binding of the cross-component references according to the architecture topology; 3) distributed management, such as communication and transaction; and 4) style-compliant reconfiguration policies, including addition, removal, and replacement of components and connectors. To be more general, the elements in our class model are based on an ACME [76] specification.

The specific architectural style-compliant model can be realized through inheritance from *RuntimeArchitecture*. Developers can also derive and tailor their own architecture class from the existing classes to best fit their application needs. For example, with the support provided in RuntimeArchitecture, a Java class of a simple Master/Slave style can be declared as illustrated in Listing 2.1.

The component and connector information gets initialized during composition time, and, thus, the runtime architectural object can be instantiated automatically based on the class template. Besides the structural topology, it also includes the style-specific coordination logic, as well as, the constraints. For example, as shown in Listing 2.1, a weak typed method invokeOnSlaves is provided to redirect calls from the master to the slaves based on the binding relationship from the connector.

Listing 2.1: Master/slave style class model illustration

```
public class MasterSlaveArch extends RuntimeArchitecture implements
    ISlave, IMaster{
    //constituent entities like components, connectors ...
    IMaster master;
    Vector slaves;
    Port drivenPort; ...
    //constructors
      public MSArch(ArchConfig ac){...}
      ...
    //methods declared in ISlave
    //for slaves to get tasks from the master
      public Object invokeOnSlaves(Method m, object[] params) throws
        Exception {...}
    //other methods declared
      ...
    //implementations for dynamic reconfiguration
      public void addSlave(ISlave s){...}
      public void removeSlave(ISlave s){...}
      public void addMaster(IMaster m){...}
      ...
}
```

The dynamic reconfigurations can be naturally modeled as the behavior of the architecture object through class methods. Clearly, these reconfigurations must be foreseen in the design phase and integrated within the provided templates. However, there are some reconfiguration requirements gradually discovered after the system is put into operation. This built-in, programmable software architecture object offers an edge in this case. Since in our approach the dynamic reconfigurations are regarded as the behavior of the architectural object, the unanticipated reconfigurations are new behaviors of the evolved object. A new class model, which inherits the provided architecture class, can be defined and dynamically loaded into the running application to replace the old one.

2.3.2 Reflective interaction

In our approach, the runtime software architecture does not only contain structural topology information but also the coordination logic. Through the reflection mechanism provided by the programming languages, the runtime architecture object is transparent to the involved components. The object-oriented paradigm of method invocation object.method() now is redirected and the architecture object reinterprets the reference.

For example, component *comp* invokes a method *m1* of component *comp1*. Normally, the code statement is *comp1.m1(args)*. Without loss of generality, we assume *comp1* is planned to be replaced by another component *comp2*, which offers a similar function but with a better implementation of non-functional properties. To update the system, the reference is changed from *comp1.m1* to *comp2.m2* and perhaps accompanied by the modification of arguments, if necessary. It is difficult—if not impossible—to conduct the dynamic evolution without shutting down the system. In our approach, because the functional component and connector descriptions are properties of the architecture object, this object intercepts the invocations among these components based on the connector's binding information. In case of evolution, the *comp2* description will replace *comp1* and the port-role binding information will be updated to the architecture object. By dynamic class loading, the updated architecture object will redirect the invocation to the method *m2* in *comp2*. In this process, not only can the dynamic evolution be realized, but also the computation states during the evolution can be restored later through utilities such as serialization or logging.

There is another benefit by leveraging runtime reflection. As the functional component can be discovered (such as services in the web) and composed into an application, by weaving their descriptions into the architecture object, the object interception process is transparent to the involved components. Despite the fact that the introduction of reflection requires more computational resources, the approach offers higher transparency and flexibility which is more desirable in dynamic adaptation.

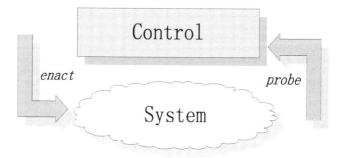

Figure 2.5: Abstraction process for self-adaptation

2.3.3 Discussions

Almost all external adaptation approaches can be generalized into a feedback-loop-like model illustrated in Figure 2.5. Systems first probe the changes from the environment, or requirement profiles defined by users, through some monitoring utilities, and then deliver the changes to the control module. The control module analyzes the information and adapts the system, if necessary. Generally, the adaptation process can be divided into four phases: occurrence of the changes, acquisition of the changes, analysis, and reaction. Changes mainly originate from the following three possible sources. *Requirement changes*: customers or users may change their requirements as time goes by. In some extreme cases, even the customers are not clear about their exact requirements at the very beginning; *operating context changes*: it includes the changes arising from hardware platforms, network conditions, etc. Nowadays computing and network infrastructure have been diversified into various forms, which are interconnected; *maintenance needs*: it includes the efforts applied to correct systems' faults, and some functional components may be required to be updated.

There are several means to observe and collect the changes. The most primitive is through manual input. The other possibility is through some physical and software sensors to capture the changes automatically. After obtaining the changes, the adaptation module needs to analyze the acquired information through the adaptation logic. There are mainly three classification criteria for adaptation logic. By the first criterion, the adaptation logic is divided into the categories of the *manual* and the *automatic*. As their names suggest, some adaptation actions are conducted through man-power, while the others are conducted through automatic tools. By the second criterion, the adaptation logic is classified as the *external* and the *internal*. External adaptation separates the control concern from the business concern and functions as a standalone module, while the internal adaptation approach is to embed and mix the logic closely with the business concern. Obviously, for complex systems, the separation between adaptation concerns and business concerns is necessary. On one hand, the functional part of the applications tangled with the adaptation part can be quite complicated; on the other hand, it is easier to make changes to the adaptation part without affecting the business logic, because it provides better reusability and controllability. The third classification criterion is from the perspective of loading time. Some adaptation mechanisms are predefined and loaded before running, while other mechanisms can be loaded dynamically at runtime. The latter category is more flexible and can support extension when necessary.

The explicit context modeling and the internal runtime architecture are two salient features of our conceptual framework. In the framework, components and connectors belong to the system part. Probes and gauges are monitoring utilities, while explicit context reasoning and reacting parts constitute the control part. The runtime architecture is the coordination entity. The control commands will first be directed to the architectural object and cause its adaptation. In turn, the updated architectural object will reinterpret the component invoking to realize the adaptation behavior. Because software architecture offers a systematic view of the application, the approach can offer an adequate generality for software adaptation in the open environment.

2.4 Related Techniques for Self-Adaptation

In this section, we will present some supporting techniques that can help enable self-adaptation. Generally speaking, software adaptation has four fundamental elements, *i.e.*, **context**, **sensing**, **decision-making**, and **adaptation** (dynamic evolution). These four elements form a closed loop and guide the self-adaptation process.

Along these four dimensions, Figure 2.6 gives a comparison framework for techniques during the adaptation process. Adaptation is a kind of software behavior, and multiple approaches can provide this functionality. Context is an abstraction of the internal and external environment of the running software. It includes not only the static information, but also the dynamic runtime information and other non-functional requirements. The context information acquisition is a fundamental requirement for the adaptation which provides the premise. The corresponding research on context-awareness is increasing recently, especially in the areas of pervasive computing and mobile computing. However, in the area of software adaptation, there is relatively less research on explicit modeling of the context. We argue that context should be regarded as a first-class entity and an explicit is essential for self-adaptation in the open environment. In this way, it can provide a uniform semantic foundation for the following sensing and evolution. Sensing is a transformation part which shifts the emphasis from the external environment to the internal software. By distilling interesting information from the context, the software can use multiple mechanisms to observe the environments and react accordingly. For example, the Rainbow system [75] proposed a probe-

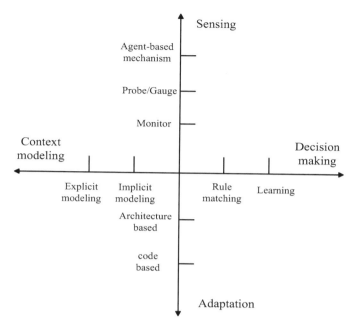

Figure 2.6: Comparison framework for adaptation process

gauge based approach to observe the context. In the aspect of decision-making, techniques in early stages mainly leverage straightforward rule matching mechanisms. Recently a number of systems employ more complex techniques, such as knowledge-based reasoning and reinforced learning, etc. The architecture-based approach is one of the dominate approaches to enable adaptation since it has many advantages in an open environment. There exist other code-level techniques, such as the dynamic translator based approach [112].

Jim Dowling *et al.* at Trinity College Dublin proposed a *K-component frame-work* to enable adaptation for decentralized applications [57]. The framework separates the concerns of the adaptation logic and the business logic. In the adaptation part, the configuration graph is employed to specify the architecture meta-model, and the reconfiguration is modeled by the graph transformation. Therefore, the adaptation is represented by the reconfiguration of the architecture. Architectural reflection is defined as being concerned with the observation and manipulation of the configuration graph of software architecture and its constituent vertices and edges. For the local component, it has a partial view of the whole architecture. The monitor module senses the context information, and the decision-making module uses reinforced learning techniques to analyze the information. In the business part, K-component framework relies on the component composition mechanisms. It uses K-IDL to specify the interfaces. The adaptation of this framework mainly denotes the substitution of new components. However, as the framework does not address the runtime state management of these components, it limits the adaptability and application of this approach.

Rainbow is an adaptation infrastructure proposed by Garlan *et al.* at Carnegie Mellon University [75]. It uses an external runtime architecture model to monitor and control the running application. An additional configuration manager is needed to map the external model to the running application. The model contains three layers, *i.e.*, the architecture layer, the translation layer, and the system layer. Raw data retrieved from the system layer will be interpreted in the translator layer. After the interpretation, the data will be understood by the architecture layer. The adaptation logic stays in the architecture layer. It will analyze and evaluate the contextual information according to current status and predefined policies. Then the adaptation decisions will be interpreted in the translator layer, and act upon the system layer. During this process, we can observe that the translator layer is usually application-specific and hard to implement.

Taylor *et al.* at the University of California, Irvine, developed an architecture-based adaptation framework: *ArchStudio* [48]. In the early stages, they proposed an architecture description language *C2* to describe the architectural styles and patterns. In later versions of ArchStudio, they further added other XML-based platform-independent languages, such as xADL, and the knowledge-based adaptation management modules [79]. Because C2 is a dynamic architecture description language, it supports multiple kinds of adaptations, such as, connectors addition/removal and components addition/removal. However, it still employs the external description approach, and a specific management module has to be designed to connect the ADLs to the corresponding runtime architecture. Besides, ArchStudio does not give enough attention to the explicit context modeling.

The EU-funded project *Arch-ware* is another representative work towards architecture-based self-adaptation. It applies an innovative approach to the architecture-centric model-driven engineering of software systems that sets the "ability to evolve" as its central characteristic. Evolution arises in response to changes of requirements, as well as, to runtime feedback [135]. The most salient feature of this project is that it tries to extend π calculus to model the architecture-centric adaptation. Besides, it proposes a set of architecture description, analysis, and refinement languages. The language set covers the scope from the architectural definition to the architectural evolution. The work has a sound theory foundation and corresponding tool support. However it lacks the direct support for the explicit modeling of the context information.

Table 2.1 summarizes the differences of the above techniques.

2.5　Summary

In this chapter, we have proposed a conceptual adaptation framework which mainly consists of the context model, the internal runtime architecture mecha-

Table 2.1: Comparison of selected adaptation projects

	Context Modeling	Sensing	Decision Making	Adaptation
Rainbow	Implicit	Probe-Gauge	Event Matching	External ADL (ACME)
K-Component	Implicit (K-IDL for states)	Monitor	Reinforced Learning	Architecture-Based Meta-Model
ArchStudio	Implicit	Bus-sensor	Knowledge-Based	xADL/C2 Based ADL
ArchWare	Implicit	Monitor	Event Matching	Model-Driven

nism, and other supporting assurance mechanisms. Based on the framework, we can identify the requirements of software adaptation and incorporate the characteristics of an open environment into the framework. Despite its simplicity, the framework lays a foundation to study software adaptation in an open environment. Compared with other framework and models, ours features:

■ More emphasis is put on the role that the context plays in self-adaptation. Incorporating the architectural and environmental information offers a comprehensive knowledge basis for adaptation.

■ Different kinds of components are supported. The framework covers the adaptation of standalone component composed systems, such as, services and components in traditional senses.

Chapter 3

Context Modeling

CONTENTS

Context-awareness is the basis for efficient software adaptation. An adequate context model can greatly facilitate the adaptation process. In an open and dynamic environment, context becomes much more complex than its counterpart in the closed and static environment, and plays a crucial role in the system's adaptation process [119]. Essentially, it requires the description of the environment and the software's internal states. The context information is provided as a knowledge base for the adaptation decision-making. For architecture-based adaptation systems, usually, there is a specific utility module to monitor the architectural models and to link the models to the system's implementation. The changes from the environment cause the reconfiguration of the architectural models, and the models' reconfiguration further influences the behavior of the software systems.

3.1 Overview of Context Modeling

Due to the importance of context in distributed systems, context modeling is an active research subject in many areas, such as pervasive computing and autonomic computing. Context modeling techniques are vastly investigated in these fields. Strang and Linnhoff-Popien observed the various representations of context information in the physical world and summarized the most relevant context modeling approaches according to the data structures used for representing and exchanging contextual information in their respective systems [166]. The six representative categories are *key-value models, markup scheme models, graphical models, object oriented models, logic based models,* and *ontology based models.* These models are all widely used, but have different abilities to support distributed composition, partial validation, richness, and quality of information, etc.

Key-value models These models represent the simplest data structure for context modeling. They are frequently used in various service frameworks, where the key-value pairs are used to describe the capabilities of a service. Service discovery is then applied by using matching algorithms which use these key-value pairs. Key-value models are lightweight and easy to manage and understand. But they lack the capabilities to describe more complex context information and the relations between the entities.

Markup scheme models All markup based models use a hierarchical data structure consisting of markup tags with attributes and content. Profiles represent typical markup-scheme models. Typical examples of such profiles are the Composite Capabilities/Preference Profile (CC/PP) and User Agent Profile (UAProf), which are encoded in RDF/S. Markup scheme models, such as comprehensive structured context profiles [94], use a hierarchical data structure consisting of markup tags with attributes and content [166]. Usually the accompanied schemes support the type checking and validation. But the incompleteness and ambiguity have to be handled on the application level.

Graphical models Examples of graphical models include unified modeling language (UML) and object-role model (ORM) [95]. Various approaches exist where contextual aspects are modeled by UML, or its extension ORM. Due to the inherent advantage of graphical representation, graphical models are user-friendly and can express meta-context information, but they lack a formal basis and are weak in partial validation.

Object oriented models Object-oriented models [160] employ the ideas of objects, such as encapsulation and inheritance, which enables the full power and advantages of object orientation. Existing approaches use various objects to represent different context types (such as temperature, location, etc.), and encapsulate the details of context processing and representation. Accessing the context and the context processing logic is provided by well-defined interfaces. This approach supports distributed composition and partial validation, but it lacks a formal reasoning mechanism to derive implicit information.

Logic based models Logic based models advocate using formal logic as a unified framework to specify context and related rules [124]. They have a high degree of formality. Typically, facts, expressions, and rules are used to define a context model. A logic based system is then used to manage the aforementioned terms and allows the addition, update, or removal of new facts. The inference (also called reasoning) process can be used to derive new facts based on existing rules in the systems. The contextual information needs to be represented formally as facts. As this model requires high expertise in logics, its application domain is limited. Besides, without partial validation, the specification of the contextual knowledge is quite error-prone [166].

Ontology based models Ontologies represent a description of the concepts and relationships. Therefore, ontologies are a very promising instrument for modeling contextual information due to their high and formal expressiveness and the possibilities for applying ontology reasoning techniques. Various context-aware frameworks use ontologies as underlying context models. Ontology-based models provide a set of supports for distributed knowledge composition and sharing. Besides, the description logic can facilitate reasoning [85]. Due to its expressing power and the formal foundation, ontology becomes a promising instrument to describe entities and interrelations.

The above list gives some representative context modeling approaches. Each approach has its own usage in different cases. These models are developed in pervasive computing or autonomic computing areas and presented as general alternative tools. Regarding context as a synonym of environment, these techniques only focus on the external circumstances of the application. In open environments, as the semantics of context changes, these tools have to be tailored or extended.

Aside from these models, some researchers also work from the theoretical perspective, such as Context UNITY. Context UNITY is a formal model and notation for expressing quintessential aspects of context-aware computations and existential quantifications. Furthermore, context UNITY treats context in a manner that is relevant to the specific needs of an individual application and promotes an approach to context maintenance that is transparent to the application [152].

The conclusion of the evaluation presented in the survey by Strang and Linnhoff-Popien [166], based on the six requirements, shows that ontologies are the most expressive models and fulfill most of their requirements. For the selective ontology models introduced above, they mostly focus on the domain knowledge in the pervasive computing environment. To extend the context models in an open environment, which has a broader coverage, the domain concepts have to be reconsidered.

3.2   Representative Ontology Models

In this section, we introduce some representative ontology models, including CoBrA [40], GLOSS [53], ASC [167], and SOCAM [86]. All of these models attempt to build domain ontologies for context-aware computing, but they have different characteristics [188].

■   The Context Broker Architecture (CoBrA) is an agent based architecture for supporting context-aware computing in intelligent spaces. CoBrA provides a means of acquiring, maintaining, and reasoning about the context. Central to the architecture is the presence of an intelligent context broker that maintains and manages a shared contextual model on behalf of a community of agents. The context broker has four main functional components, *i.e.*, context knowledge base, context reasoning engine, context acquisition module, and privacy management module. Contextual information in CoBrA is expressed by a set of ontologies called COBRA-ONT that is implemented in OWL. The ontology defines a vocabulary for describing people, agents, places, and presenting events for supporting an intelligent environment. It also defines a set of properties and relationships that are associated with these basic concepts. The proposed ontology is categorized into four distinctive but related themes. The first is the concepts that define physical places and their associated spatial relations. The second is the concepts that define agents. The third is the concepts that describe the location contexts of an agent in the environment, and the last theme is about the concepts that describe the activity contexts of an agent.

■ GLObal Smart Space (GLOSS) is a framework that describes a universe of discourse for understanding global smart spaces. The key concepts are people, artifacts, and places. Central to the GLOSS is the provision of location-aware services that detect, convey, store, and exploit location information. The GLOSS ontologies describe a small set of concepts for a universe of discourse. These concepts provide the precise understanding of how services are used and how users interleave various contexts at run time. GLOSS permits the dynamic rearrangement of low-level interconnection topologies and the components they connect. Service evolution is driven by the GLOSS ontologies.

■ Aspect-Scale-Context (ASC) model has three core concepts, *i.e.*, aspect, scale, and context information. An aspect is a classification, symbol, or value range, whose subsets are a superset of all reachable states, grouped in one or more related dimensions called scales. Contextual information is used to characterize entities relevant to a specific task in the related aspects. Context ontology language (CoOL) is developed to describe contextual facts and contextual interrelationships in a precise and traceable manner. CoOL contains two subsets: the CoOL Core, which projects ASC model into various common ontology languages such as OWL. Another subset is CoOL Integration, which is a collection of schema and protocol extensions, as well as, common sub-concepts of ASC. As context bindings are useful to link service parameters to well defined aspects and scales, the framework can fit well into the modeling of web service architecture and its protocols.

■ Service-Oriented Context-Aware Middleware (SOCAM) is a middleware architecture that aims at helping application programmers to build context-aware services more efficiently. The core components of the architecture include context provider, context interpreter, context-aware services, and service location services. To facilitate context description, the CONtext ONtology (CONON) is proposed. The ontologies include a common upper ontology for the general concepts in pervasive computing, such as person, location, and computing entity. The context model supports multiple semantic contextual representations such as classification, dependency, quality of the context, and reasoning. The CONON ontologies help to share a common understanding of the structure of contextual information from users, devices, and services so as to support semantic interoperability and reuse of domain knowledge. They also support efficient reasoning mechanisms so as to check the consistency of context and deduce higher-level, implicit context from raw context [86].

3.3   Ontology-Based Context Modeling

Ontology is a concept in philosophy concerned with the nature of existence. For knowledge-based systems, the term means a specification of a representational vocabulary for a shared domain of discourse — definitions of classes, relations, functions, and other objects, as in software literatures, what "exists" is exactly that which can be represented [85]. In [176], Uschold and Gruninger summarized the usage space of ontology and classified it into three categories.

1. Communication. Ontologies reduce conceptual and terminological confusion by providing a unified framework within an organization. Thus, ontologies enable shared understanding and communication between people with different needs and viewpoints arising from their particular contexts.

2. Inter-Operability. Ontologies can address the issue of inter-operability. A major theme for the use of ontologies in domains such as enterprise modeling and multi-agent architectures is the creation of an integrating environment for different software tools.

3. Systems Engineering. This point focuses on the role that ontologies play in the operation of software systems. A shared understanding of the problem and the task at hand can assist in the specification of software systems. Due to the common understanding and unambiguity of the specifications, ontologies can help improve systems reliability and reusability.

As ontologies describe a shared domain of discourse including objects, relations, etc., **ontology-based context modeling** can help eliminate unambiguity and facilitate sharing and reuse. Therefore, this offers a cutting edge in modeling the complexity of context in an open environment.

The essence of ontology is to capture the natural features of entities and their relations. Due to the complexity of the realities, entities have different levels. Some knowledge is general, while some others are relatively more specific. Therefore, ontologies can be classified into four layers according to their generality, *i.e.*,

■ Representational ontology, which stays at the meta-level and provides the most basic primitives for upper layers, *e.g.*, the frames, the slots, etc.

■ Generic ontology, which is more specific than the representation layer. It has concepts such as time, space, etc. Domain ontology describes domain-specific vocabularies,

■ Domain ontology, and

■ Application ontology, which can be regarded as an instance of the domain ontology. The lower layers provide better reusability and the higher layers provide better applicability.

Currently, the dominate markup language to create ontology is *Web Ontology Language* (OWL). It is a standard ontology description language recommended by W3C in semantic web. The OWL language is an extension of the DAML+OIL web ontology language. DAML+OIL was developed by a group jointly funded by the US Defense Advanced Research Projects Agency (DARPA) under the DAML program and European Union's IST funding project. OWL has three variants with different expressiveness, *i.e.*, OWL-Lite, OWL-DL, and OWL-Full.

There are basically two ways for constructing contextual ontologies. The first way is to design a comprehensive set of ontology models for the target subject. This can be contributed by the discipline of ontology engineering. Some notable examples are General Formal Ontologies (GFO)[1] and OpenCyc[2]. These ontology models are general and standardized, and thus could be widely used across different domains. The second way is specific domain or feature oriented instead of providing a general complete ontology model. Since the first approach requires a clear boundary of the target domain studied and has many irrelevant concepts for self-adaptation and these concepts bring unnecessary complexity, we adopt the second approach to build the context ontology. We use the following example to illustrate our ontology modeling process.

In an online ticket reservation system, one of our concerns regarding the system's non-functional properties is responsivity. The related context information is user's experience (*UserExperience*) which denotes the time interval between request and reply, and network latency (*NetworkLatency*) which denotes the network conditions connecting the user and the server-side. The snippet of context ontology is shown in Figure 3.1.

Listing 3.1: Context ontology illustration

```
<owl:Class  rdf:ID="UserExperience">
  <rdfs:subClassOf  rdf:resource="#Context"/>
  </owl:Class>
<owl:DatatypeProperty  rdf:ID="interval">
  <rdfs:range  rdf:resource="http://www.w3.org/2001/XMLSchema#int"/>
  <rdfs:domain  rdf:resource="#UserExperience"/>
  </owl:DatatypeProperty>
<owl:Class  rdf:ID="NetworkLatency">
  <rdfs:subClassOf>
    <owl:Class  rdf:ID="Context"/>
  </rdfs:subClassOf>
  </owl:Class>
<owl:DatatypeProperty  rdf:ID="value">
  <rdfs:range  rdf:resource="http://www.w3.org/2001/XMLSchema#int"/>
  <rdfs:domain  rdf:resource="#NetworkLatency"/>
  </owl:DatatypeProperty>
```

[1] http://www.onto-med.de/ontologies/gfo/
[2] http://www.onto-med.de/ontologies/gfo/

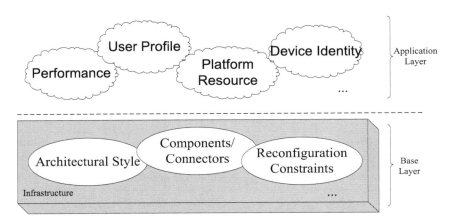

Figure 3.1: Two-layer context classification

3.4   Ontology Models for Architecture Domain

Existing work of ontology-based context modeling emphasizes the physical aspect of environment modeling, such as personal identity, location, activity, and computational entity [183]. In this sense, context is equal to the external environment. But as analyzed previously, some internal information about the software itself, such as the architecture and the state, can also be part of the context. Extending the ontology model into the architectural domain is necessary. In this way, the architectural knowledge can be represented, and this knowledge can be used to facilitate adaptation. For example, a master/slave architectural style application has two masters and several slave components. Usually two master components can guarantee that the system responds quickly. If the client still reports low response, the adaptation inference module can decide that the bottleneck is not in the server side, but perhaps in the poor network condition. To enumerate another example, in the case of component level migration with the user, the migration decision should be based on the component properties and their interaction conditions with other components. It also requires the internal structure information of the application itself.

We classify the context into two layers: the upper part is the *application layer*, and the lower part is the *architectural or base layer*. Figure 3.1 illustrates the classification. One of the benefits of this classification is that it gives a clear view of the separation of concerns between infrastructure and upper applications. In this way, infrastructure designers can focus on the base layer, and just offer a general platform for the application layer, although upper context information may vary from application to application. As shown in Figure 3.1, the context information stretches from the environment to the system. To model the context entities, we present a set of ontology models which depicts both the runtime system and execution environment. Roughly, these ontology models can be divided into four sub-models: environment ontology, architecture ontology, interaction ontology, and transformation ontology.

Environment ontology mainly consists of the physical devices, platforms, and environmental properties, such as temperature, location, and noisiness. Interaction ontology addresses the aspects of interaction properties, such as users' interaction requirements, and users' operation habits. Architecture ontology includes the architecture description ontology, architecture control ontology, and architecture manipulation ontology. Because of the knowledge gap between these separate domains, there is a need for a translation to map the different terms with similar semantics. This issue is addressed by the transformation ontology. As the first two ontology models are common in literature, we will focus on the other two kinds.

The architecture ontology does not only describe the static configuration but also the dynamic evolution of the system's architecture. Moreover, it specifies the control over the evolution processes targeting the requirement goals. Consequently, there are three architecture ontologies respectively: architecture description ontology (ADO), architecture manipulation ontology (AMO), and architecture control ontology (ACO). ADO corresponds to the general architecture description languages and mainly describes the static configuration of the architectures. AMO defines the actions to operate on the architectures. ACO associates AMO with ADO by specifying when to carry out the actions in AMO to manipulate the architectures described with ADO. AMO together with ACO achieves the dynamic management function over software architecture.

Besides, ADO and AMO also support architectural styles. Architectural styles provide a terminological space for the software designers to communicate with each other and promote the reuse of the designs and the codes. To fulfill the aim, ADO and AMO are further classified into three layers, namely meta layer, style layer, and instance layer. These architectural ontologies are illustrated in Figure 3.2 which also contains some transformation ontologies (TO), such as, architecture-interaction TO and environment-architecture TO. The details of architecture ontologies and the utility ontologies are explicated in the following subsections.

3.4.1 Architecture description ontology

The design of ADO refers to ACME [76] architecture description language due to its generality. As aforementioned, ADO consists of three layers. The details of the three layers are meta layer, style layer, and instance layer.

The meta layer has seven core concepts defined, *i.e.*, *component, connector, port, role, system, representation,* and *rep-map*. These concepts correspond to the peer concepts in ACME. Briefly speaking, *component* denotes the general functional or computational entity of the model. *Connector* represents the interaction or communication part of a model. *Port* is the public interface, through which a

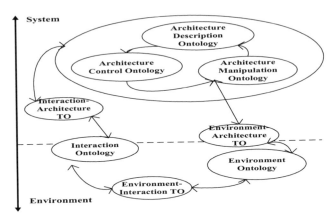

Figure 3.2: Context ontology models

component can interact with other components. Usually a component can have multiple ports. *Role* is the public interface of a connector. The relation of a *role* to *connector* is similar to that of *port* to *component*. Roles have to be bound with ports so that their host components and connectors can interact with each other. *System* is a term to denote the composed model from *components* (with ports) and *connectors* (with roles). It describes the basic configuration of a model. *Representation* is a construct to further elaborate an entity, for example, a composite component or a sub-system. *Rep-map*, *i.e.*, representation map, defines the mapping relation between the representation of an internal sub-system or other kinds of compound entities and the external entities.

The style layer of ADO defines the vocabulary for specific architecture styles. For instance, the Master/Slave style has the concepts of Master, Slave, and so on. In addition to the attributes shared by all general components, they have some special ones adhered to the Master/Slave style.

The instance layer contains the description for a particular architecture configuration, *i.e.*, the application level instances of ontologies. For example, in an online ticketing system designed following the Master/Slave style, the components of bus ticket booking (aBusTicketService), plane ticket booking (aPlaneTicketService), and train ticket booking (aTrainTicketService) are all instances of the instance of "slave" ontology, while a frontend task receiving and dispatching component could function as an instance of "master" ontology belonging to this layer.

3.4.2 *Architecture manipulation ontology*

ADO defines the structural topologies of a system, while AMO is applied to describe available reconfiguration actions for the architecture. Therefore, AMO exhibits the operational aspects. Corresponding to ADO, AMO also has three

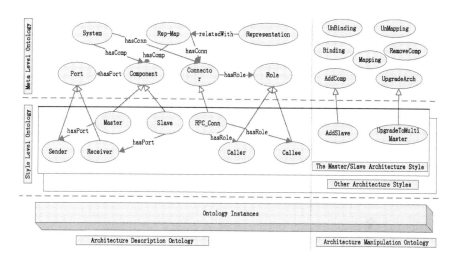

Figure 3.3: Relationship between ADO and AMO

layers. In addition, the actions in these three layers can be further classified into global actions and partial actions respectively according to their action scope.

In the meta layer there is a global action *UpgradeArch* which is defined as an upgrading function over the architecture. This action is a general and abstract concept that is supposed to be reified into corresponding concrete ones with respect to specific architecture styles in the style layer. There is also a set of primitive actions including *AddComp*, *RemoveComp*, *AddConn*, *RemoveConn*, *Binding* (bind ports and roles), *Unbinding*, *Mapping* (map architectural component such as a Slave to a physical component such as an EJB), and *Unmapping* which is a reverse action of *Mapping*. These primitive actions can be used individually or combined to construct compound actions.

In the style layer there are style-specific actions. For example, regarding the Master/Slave architecture style, the global action of UpgradeToEMS can be defined to conceptualize the up-grading from the common Master/Slave style to the Extended Master/Slave style. The primitive actions include *AddMaster*, *AddSlave*, and so on. These actions define the architecture style-specific actions.

In the instance layer, concrete actions specific to applications are described. The ontologies in this layer are left to application developers to define. They can reuse the infrastructure ontologies in the two basic layers and customize based on their requirements.

Based on the above explanation, we can observe that ADO and AMO share a similar layered structure. Figure 3.3 illustrates the relationship between ADO and AMO.

3.4.3 Architecture control ontology

ACO also describes architecture dynamics. But different from AMO, it emphasizes the conditions and consequences of related actions. For example, when an action or a sequence of actions defined in AMO should be carried out and what the intended consequence is. It is often closely related with some essential properties for a certain architectural style. For instance, the rationale behind a Master/Slave architecture is that the performance capacity can be adjusted with addition/removal of slaves. Information conveyed in ACO is in the form of triples of Condition-Operation-Consequence. *Condition* is the prerequisite of carrying out an Operation. *Operation* could be either a single action or a combination of multiple actions defined in AMO. *Consequence* declares the effect of the *Operation*. To better capture the semantics of ACO, instead of using RDF based XML formats, we use the following style to present ACO.

The primitive action shown in Listing 3.2 describes that an additional slave component is needed when every existent slave component's load reaches its capacity limit while the total load of the whole system is still expected to increase. After a new slave component is added successfully, the load is balanced and all slave components can be transitioned into the status of "Not Full" again. The second example below illustrates a combination of a global action and a primitive one. It depicts a solution, that is, to upgrade the common Master/Slave style into an EMS style, for the system performance issue when even the load of Master comes close to its capacity limit. The EMS style has one more Master than the original one. With the additional Master, the relation between Master and Slave as well as other system properties need to be reconsidered. Thus, the action UpgradeToEMS has to be executed in addition to the AddMaster action as illustrated in Listing 3.3.

Listing 3.2: ACO action illustration

```
Condition: NOT satisfied(System.performance) AND full(Slave.capacity)
Operation:AddSlave
Consequence: NOT full(Master.capacity)
```

Listing 3.3: ACO action illustration 2

```
Condition: NOT satisfied(System.performance) AND full(Master.capacity)
Operation:UpgradeToEMS, AddMaster
Consequence: NOT full(Master.capacity)
```

These ontologies can effectively describe the concepts within their respective domains, and they provide a consistent representation for both internal architectural information of software systems and external information of environments. Nevertheless, we can observe that there are still semantic gaps between above ontologies. Therefore, additional ontologies are developed to glue together concepts in different ontologies which function like ontology mapping. The mapping can be trivial for essentially coincident but differently represented concepts. It could also be rather complicated with computations for relevant but different concepts. Accordingly, we have designed three kinds of transformation ontologies, *i.e.*, interaction-environment ontology, interaction-architecture ontology, and environment ontology. As for our example of the online ticketing system, *Responsivity_Goal* in the interaction ontology expresses the user requirement for the system's responsivity with the property of rate while architecture ontology defines a *performance property* of system to label the system capability. At this scenario, *Responsivity_Goal.rate* is essentially coincident to the System.performance. And such a coincidence is going to be caught with the interaction-architecture ontology. Similarly, the real responding time is the difference between the user experienced delay and the network latency. The conversion from UserExperience.interval and NetworkLatency.value to *Responsivity_Goal.rate* is done with the interaction-environment ontology.

As to ontology editors, several tools can be employed. Among these, we choose web ontology language (OWL) for its generality [7]. OWL is a semantic markup language for publishing and sharing ontology proposed by W3C's Web Ontology Working Group. It is developed as a vocabulary extension of Resource Description Framework (RDF) and derived from the DAML+OIL [97]. OWL follows the XML syntax style and has the advantage of platform independence.

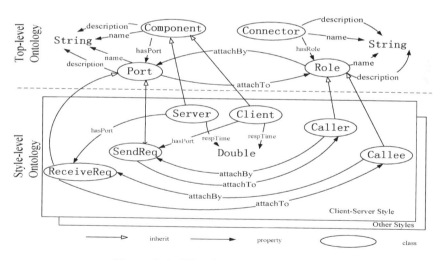

Figure 3.4: Client/server style ontology

Figure 3.4 illustrates a client/server architectural style ontology. *Ports, components, roles,* and *connectors* are all key general elements of software architecture, while *server, client, caller,* and *callee* are terms of a specific architectural style. This relationship and the interactions between the entities are also described in this illustration. The corresponding OWL-encoded description for the *Component* part of the above sample ontology is shown in Listing 3.4.

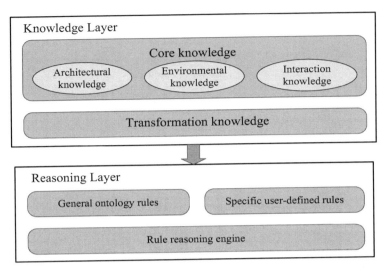

Figure 3.5: The context model

Listing 3.4: OWL description snippet illustration

```
<owl:Class rdf:ID="Component">
 <rdfs:comment>This is component description</rdfs:comment>
 <rdfs:subClassOf>
  <owl:Restriction>
   <owl:onProperty>
    <owl:ObjectProperty rdf:ID="hasPort"/>
    <owl:cardinality rdf:datatype=http://www.w3.org/2001/XMLSchema#int>1
    </owl:cardinality>
    ...
 </rdfs:subClassOf>
 <rdfs:subClassOf>
  <owl:Restriction>
   <owl:onProperty>
    <owl:DatatypeProperty rdf:ID="description">
    <rdfs:range rdf:resource="http://www.w3.org/2001/XMLSchema#string/>
    ...
</owl:Class>
```

The above ontology models provide a comprehensive support for describing context. However, to achieve the desired adaptability by utilizing the knowledge, appropriate transformation facilities should be built up. Since these models employ OWL/RDF as the underlying description language, the facilities are naturally a set of reasoning engines based on the language. We have two main kinds of rules. The first kind is the general OWL ontology rule, which is used to deduce some implicit knowledge from the explicit ontology models. The other kind is the application-specific rule. This kind is customizable and not limited with description logic or first-order logic. It can be like event-condition-action tuples or even specific term rewriting techniques, so that they are able to describe adaptive behaviors directly and efficiently.

According to the above analysis, our context model has two basic layers as illustrated in Figure 3.5, *i.e.*, knowledge layer and reasoning layer. In knowledge layer, it contains core knowledge and transformation knowledge. The core knowledge further contains architectural knowledge, environmental knowledge, and interaction knowledge. The reasoning layer mainly contains rule reasoning engines which support general OWL rules and application-specific, user-defined rules.

3.4.4 Discussion

The context ontology and architecture ontology provides a uniform representation of each domain, but there is still some semantic gaps between them. This is due to the fact that some concepts are essentially semantic-equivalent but have different representations in their own domain. Some utility ontologies are specially designed to bridge such gaps, such as, interaction ontology and transformation ontology. Interaction ontology mainly denotes the ontology models defined for interaction requirements from end users. Transformation ontology targets for the semantic translation among the ontologies from different domains. Accordingly, there is environment-architecture transformation ontology, interaction-architecture ontology, and environment-interaction ontology, which are shown in Figure 3.2. For example, in an online ticket reservation system, from the users' perspective, the concept of responsivity might be described by the *rate* of *Responsivity_Goal* from interaction ontology, while in architecture ontology, the same concept is described by *performance* of *system*. In this way, the two concepts essentially describe the same target. Thus, the transformation ontology can semantically bridge the gap between *Reponsivity_Goal.rate* and *System.performance*.

3.5 Summary

In an open environment where multiple forms of devices and networks co-exist, it's essential to provide a uniform representation of the underlying context information, based on which such information could be utilized by the self-adaptation. In this chapter, we present our ontology based context modeling approach. The most salient feature is to extend context models into the architecture domain. In this sense, the contextual information also covers the internal structural information of underlying software systems. Since software adaptation requires the knowledge from both sides, this extension can provide better support for the adaptation process. In addition, to bridge the semantic gap between the different domains, some transformation ontologies are also required. These ontologies together constitute the knowledge base for the self-adaptation.

Chapter 4

Implementation and Case Study

CONTENTS

*This chapter is modified and updated from Computing, 96(8), 2014:725-747, with permission from Springer Science+Business Media.

In this chapter, we present our implementation for the support of the afore-mentioned techniques, as well as, a case study to illustrate the prototype. In light of the broad application of component based systems in practice, we leverage the coarse-grained components and services as the underlying software compo-nent model, and the simple object access protocol (SOAP), agent, and Remote Method Invocation (RMI) technology as the underlying communication utilities. Based on the framework in Chapter 2, we have implemented a middleware **MAC-ng** to facilitate the development of self-adaptive software applications. MAC-ng supports the context modeling, component composition, and architecture based adaptation, etc. In the sequel, we will first give a structural overview of the mid-dleware design. Then we proceed to describe some key modules of MAC-ng followed by a performance analysis on the overhead introduced by the ontology based reasoning and reflection. In the end, we will present a case study on the water management information system to demonstrate our prototype.

4.1 Structural Overview

MAC-ng attempts to provide a comprehensive support for the development of component-based self-adaptive software systems. The overview of the middle-ware platform is shown in Figure 4.1.

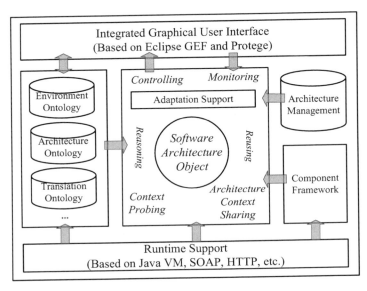

Figure 4.1: Overview of the supporting platform

The conceptual view of the system consists of six main parts: runtime support, integrated graphical user interfaces, ontology management, software architecture management, runtime software architecture objects, and component framework. Their basic functionalities are described as follows.

- The underlying supporting platform provides infrastructural facilities, such as hardware, operating system, and network connectivity.

- The integrated graphical user interface part provides the interface with the end users, and also displays all the necessary information from development to the runtime management.

- The software architecture object part takes charge of maintaining the real-time architecture information, including system topology, component properties, and connector/component composition information, etc. It is responsible for expressing and enacting the potential dynamic adaptation at runtime. Moreover, it also closely connects with other parts of the platform, for example, the context ontology part and user interface part. It is the centerpiece of the platform. There are some supporting adaptation utilities in this part, which will be discussed in Section 4.1.1.

- The software architecture management part defines a set of different composition mechanisms for the components and connectors and also the reconfiguration rules for different architecture styles. The software architecture management also implements the ontology-based adaptation executing engine for self-adaptive software and collects interested runtime information for the monitors.

- The component framework offers the essential functionalities for the component management, for example, publishing, retrieving, and composition of underlying components. This part will be further elaborated in Section 4.1.2.

- The ontology part mainly consists of the set of context ontologies introduced in Chapter 3. These ontologies, for example, environment ontologies, architecture ontologies, and transformation ontologies, are responsible for the context modeling and probing, digesting the raw data, and distilling them into some useful information for potential consumers. The storage and persistency mechanisms are discussed in Section 4.1.3.

4.1.1   Adaptation support

Adaptation support collaborates closely with the architecture object at runtime. It treats decision factors and reasoning rules as input and system adaptation actions as output and is made of three main modules: a decision factors management module, a self-adaptation logics management module, and a self-adaptation decision module.

■ The decision factors management module is in charge of maintaining all decision factors of the whole system. This module receives decision factors from the monitors. Ontology has prescribed a vocabulary to describe decision factors. The data will be stored in some database according to the properties of decision factors. Here we have designed two kinds of decision factor libraries: the standard and the extended library. The standard library stores decision factors expressed in standard RDF tuples. The extended library stores other factors expressed in extended RDF. The extended library exists for processing those complex decision factors that cannot be easily expressed by standard RDF tuples.

■ The self-adaptation logic management module maintains reasoning rules for adaptation strategies. The module receives user requirement specifications and validates these requirements with rule patterns. A rule pattern is used to ensure that input rules follow the pre-defined syntax. We classify the rules into *permanent* rules and *on-demand* rules. Permanent rules participate in the decision-making process every time; on-demand rules only take effect after their related factors change. This classification helps to reduce the scale of rule sets and improves performance of reasoning.

■ The self-adaptation decision module includes a pre-processor and a decision engine. To guarantee the generality and efficiency of decision engine, it only supports reasoning standard RDF tuples. To deal with data in the extended library, we designed a pre-processor. This pre-processor can transform extended RDF data into standard ones according to pre-defined processing algorithms and transfer these data to the decision engine.

4.1.2   Component framework

Component framework is a software entity that supports the components to conform to certain standards and allows instances of these components to be "plugged" into. It also establishes environmental conditions for the component instances and regulates the interaction between component instances. The objective of our component framework is to offer a consistent model hiding the low-level heterogeneous information for adaptive software development in an architectural approach. Software architecture is regarded as an explicit object and constructed as a software agent carrying the corresponding architecture information interacting with involving hosts and coordinating these physically dispersed components. We endow the component with the ability to sense the changes of the context by detecting periodically.

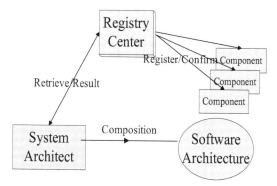

Figure 4.2: Publish/subscribe view

Any component that wants to declare itself for public use can register itself at a registry center. The registry item contains necessary descriptions, including the location, the service type, and the interfaces it provides and requires, etc. If system architects want to build applications, they just retrieve from the center and look for satisfactory components residing in various hosts. After that, those components are composed into a specific software architecture belonging to a particular architecture style. This process is illustrated in Figure 4.2.

A software component can be deployed independently and is subject to composition by third parties. In this sense, a component usually contains a unique name to identify, interfaces to access, and binary code to function. Our framework is based on Java language and XML technology, so the functional part of the component is packaged in Java class forms, while the descriptive part is delivered in XML forms. More precisely, the concrete form of the functional part can be EJB, web services, or mixed modes, and the accompanying description file is WSDL-like. Generally speaking, a component either offers an independent service or requires the collaboration of other components. However, WSDL standard does not support this. So we extend this standard to enhance its description ability, by which interface dependence can be described. Corresponding to the concepts of facet and receptacle in CORBA's CCM model in a certain degree, the notions of provide and request are introduced.

Figure 4.3 shows the basic structure of our description model. We define separate portTypes such as 'provide' and 'request' which contain corresponding operations for different component interfaces. Besides, in order to be retrieved and accessed by the clients, the name and address should also be supplied. As mentioned above, this description is not only suitable for EJB, but also compatible with web services which are illustrated in the figure.

Systems are composed of not only components but also connectors. A connector is a special type of entity. To decouple the communication from com-

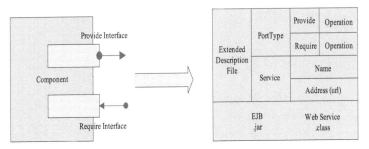

Figure 4.3: Component description model

putation from the components and localize it, necessary information should be provided, among which are the protocols, the types, and the role constraints, etc.

4.1.3 Context knowledge management

As introduced in previous chapters, we use ontology and RDF to model the context information. To store the knowledge of the context information, there are two mechanisms basically, *i.e.*, the external storage and the internal storage.

Internal storage means the runtime knowledge storage. Internal storage is not enough in urgent cases, *e.g.*, collapse or sudden power-off. To save the context information, the external storage mechanism is needed. Our context knowledge module is based on Jena [123]. Thus, the internal storage mechanism is provided by Jena. In running applications, the knowledge is represented as triple statements: $Statement := (subject, predicate, object)$.

Generally, subject is the resource name, predicate is the properties of the resource, and the object is the value of the properties.

External storage denotes the persistent mechanism for the contextual knowledge. Currently, we support two forms of persistency: the file based and the database based. For instance, OWL/RDF is a file based context persistency mechanism. Jena can retrieve information from OWL files, but it cannot transform the internal knowledge to the external OWL/RDF files. We have implemented a specific module to extend Jena's ability to support this functionality.

4.2 MAC-ng Implementation

The design aim of MAC-ng is to provide a user-friendly and comprehensive platform to facilitate the development of self-adaption applications in an open environment. It incorporates the approaches presented in previous chapters and supports the whole process of context modeling, probing, and adaptation.

MAC-ng has two main functional units. Logically, one is for distributed

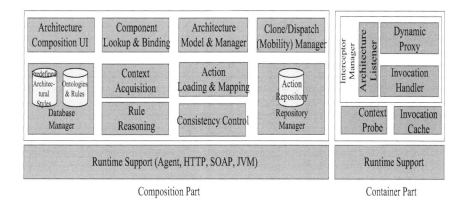

Figure 4.4: MAC-ng framework architecture

component-based application composition and context monitoring; the other is for runtime architecture listening, acting, and context probing. Physically, the composition unit is for the system developer (component assembler); the other part is distributed in the involved component containers as plug-ins. The architecture of MAC-ng is described in Figure 4.4. In distributed component based settings, the applications are composed from the third-party components and these components usually run in their separate domains. Thus, the system integrators are not necessarily the component developers. Accordingly, our approach has two parts: one is for component integrator, and the other is for component container. They function together to support dynamic adaptation of distributed component based applications.

In the component composition process, the assembler first selects a particular architectural style and designs the topology architectures of the application by drawing components and connectors through architecture composition UI. After that, the corresponding components and communication protocols are looked up and bound to the architectural components and connectors. As the interfaces of the components can be parsed through their description meta-files, the interface and invoking information are bound to the ports of the components and the roles of the connectors.

After binding the component specific information to the designed architecture, the architecture agent is generated through the template class stored in the database of architectural style classes. The clone/dispatch manager is responsible for duplicating the generated agents and dispatching them to the involved component containers.

The context related modules are responsible for context data manipulation. The context data, such as the memory usage, the CPU load, and the network condition, are sent through communication links from involved components or other places. These data are processed by the context acquisition module and are

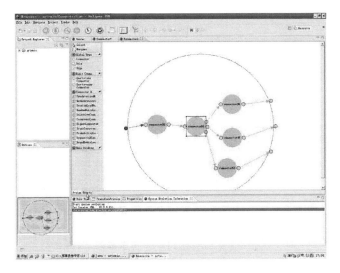

Figure 4.5: Connector composition

encoded into a unified RDF format. Then, the processed data will be sent to the interpretation and reasoning engine.

The reasoning engine will conduct the logic reasoning according to the rules and deduce whether it will issue an event signal. When a certain signal is issued from the engine, it will be transformed to a corresponding architectural configuration action. For example, when the trust value of a component drops lower than 60 percent, the 'unsafe' signal will be issued, which will lead to the detaching of this component. To fire the action, a monitor must be provided to proactively listen to the reasoning result and map it to the architectural reconfiguration. In our approach, the mapping script is encoded in a triple format: <*signal, action, parameters[]*>. *Signal* means the result issued from the reasoning engine; *action* represents the architectural configuration methods; *parameters[]* denotes the candidate architectural elements to be reconfigured. The actions will be statically checked by the conflict detection module which is explained in Chapter 6. If there are some potential conflicts detected, some warning signs will be reported to the integrator for further investigation.

Atom actions for architectural reconfiguration are: *add/removeComponent, add/removeConnector*, and *replaceComponent/Connector*. They are serialized in the action repository and can be dynamically loaded and invoked given the parameters specified by the triples. MAC-ng supports the compound composition of connectors to enable the adaptive multi-mode interaction. The GUI with the corresponding automatically generated specifications are illustrated in Figure 4.5 and Figure 4.6, respectively. The actions will effect the graphical views of the running application and regenerate a new architecture object which will be cloned and dispatched to the involved component containers.

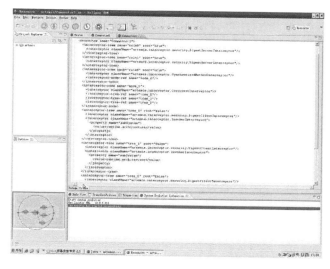

Figure 4.6: Corresponding specifications

In the component container part, the unit functions as plug-ins of the component container. It listens on a port to receive the runtime architecture agent, as the object contains the information of the component residing in this container, it can offer a local view of the whole architecture. The invocation of the component functions will be handled by the dynamic proxy, and the invocations will be redirected to the invocation handler. Therefore, the codes from the handler of architecture objects will be executed. Through dynamic proxy and class loading, this process is transparent to the caller components. The invocation cache module works with the invocation handler to preserve and recover data states during the update process of the original architecture object.

The context probe module is responsible for collecting the component host performance related information, such as the memory level and the bandwidth. Since the architecture object contains the host information of the assembler, the component's context information will be reported to the context acquisition module in the assembler side through a communication link. The runtime support offers some common utilities, such as, the reflection infrastructure, the communication, the application container, and the networking. The consistency control module supports two-level control. The first level is conducted by the adaptation rule conflict detection, and the second level is done by the architectural style checking which is implemented as a grammar directed editor in the GUI. Chapter 6 explains this part in detail.

The working system is implemented in Java by utilizing a set of open source projects and closely integrated with the Eclipse platform. To be general, the component model is the service model implemented with EJB technology. These components and associated interfaces, ports, protocols, etc. are described explicitly with WSDL language. Discovery and binding of these services are supported. We use JBoss AS as the component container and Axis as the service platform. JBoss is a micro kernel application server and based on Java Management Extension (JMX) infrastructure, which can be easily extended. JBoss supports hot deployment of components which is essential for dynamic evolution. The architecture object receiver is implemented as an mbean in JBoss and starts automatically when the JBoss AS boots. The invocation interception is realized through the Java reflection mechanism.

To support the graphical architecture composition, we build tools upon the Graphical Editing Framework (GEF) plug-ins. Developers can graphically specify the concrete configuration of the application architecture with a chosen architecture class, and bind the services discovered to the graphical components in the architecture. The system can automatically instantiate the distributed but shared architecture object. The system embeds an OWL editor, based on the APIs from protégé. A Jena reasoning engine is used to infer and trigger proper reconfiguration actions (including architecture object upgrading) with the autonomy rules, ontologies, and the context information.

4.3 Performance Analysis

In this section, we will use a simple online ticket-booking application to illustrate the concepts and conduct the performance analysis. Consider a simple distributed service-based system: there are several self-contained ticket-booking services scattered on the Internet. They sell different kinds of tickets, such as, bus tickets, train tickets, and plane tickets. The third-party service assembler, such as, a travel agency, wants to integrate these separate services and offer a single entry point to get all the ticket options for the clients' travel routes. To compose these services, another component which interacts with users and invokes separate services is needed. As the number of clients increases, the system's performance becomes poor. Customers have to wait longer to book a ticket for their travel. Now, the system assembler needs to add a backup master component to balance the load. This kind of scenario is quite common in the real world and subject to the coarse-grained, architectural adaptation. Generally, the assembler has to shut down the system and reconfigure the composition logic to evolve the system. This off-line evolution inevitably causes considerable time of the system's unavailability. As these service entities are autonomous and do not know the existence of others, the central design patterns such as publish-subscribe fail to address this problem. To facilitate this process, dynamic addition or removal of constituent entities, *i.e.*, self-adaptation is highly desirable.

4.3.1 Experimental setup

Supporting dynamic evolution introduces additional computation costs. In order to evaluate the proposed approach, we have developed the separate ticket-booking components described with WSDL (web services) and registered at a local UDDI center. These services are deployed on several PCs. These PCs have similar hardware and software configuration with Pentium 4, 2.0GHz and are connected by 1000Mbps ethernet. The component container is JBoss 3.4.2 plus Tomcat 5.5. To measure the ticket-booking server load, we use the latency between the HTTP request and response.

According to the description of the scenario, the application architecture belongs to Master/Slave style, the entry component functions as the master, and delivers the calls to the ticket-booking components. The master component has three main functions, *i.e.*, *IMRefresh*: to retrieve available ticket service information, *IMSearch*: to retrieve ticket options given the departure and destination specified by the clients, *IMBook*: to book the tickets; the slave components have corresponding three main functions, *i.e.*, *ISGetIdentity*: to return the tickets service description, *ISSearch*: to search locally the tickets information given the start, destination and the date information, *ISBook*: to book the tickets locally. The database manager system for the application is MySQL 4.1. To reduce the interference from the outside, the experiment is done in a local area network which is isolated from outside. To simulate the concurrent multiple clients' request, we write a client who issues http requests to the online ticket-booking application. The number of requests increases steadily, first 200 concurrent threads, then 400 threads, and then 600 threads. These http requests invoke the three functions (*i.e.*, refresh, search, book) randomly. When exceeding the threshold as set in Listing 4.1, it will trigger the event of adding a backup server on the fly.

The rule specified for the signal action is illustrated in Listing 4.1. This rule script means, when the response time exceeds 2500 milliseconds, it will issue an *overload* signal. The signal will be mapped to an architectural reconfiguration action. In our case the action is mapped to *AddMaster* action.

4.3.2 Performance evaluation

In order to validate the effectiveness of our approach, we have evaluated the runtime overhead and compared it with the normal runtime cost of the compo-

Listing 4.1: Rule illustration

```
@prefix ns:<http://ics.nju.edu.cn/saon#>
@prefix xsd:<http://www.w3.org/2001/XMLSchema#>
[r0:(?n ns:respondTime ?t), greaterThan(?t, '2500'^^xsd:double),
(?n ns:trustSrcNode ?srcNode),(?trust ns:trustDesNode ?desNode),
(?n ns:hasAction ?action) -> (?action ns:actName "overload"),
(?action ns:actSrcNode ?n), (?action ns:actDesNode ?n)]
```

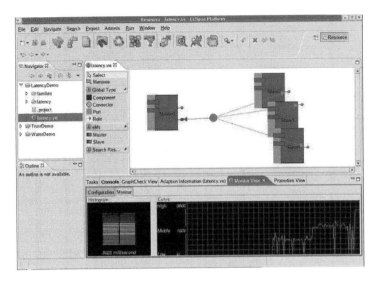

Figure 4.7: Before adaptation

Table 4.1: Time cost for architectural object generation and substitution

Object Generation	Evolved Object Generation	Substitution
103 ms	98 ms	332 ms

nent container. This overhead is mainly introduced by invocation interception (reflection). We also evaluate the influence of application's dynamic adaptation. It includes the cost of the updated architecture object generation and substitution. The average response time of the application before and after the dynamic evolution is also given. The architectural view of the application during self-adaptation is illustrated in Figure 4.7 and Figure 4.8.

To evaluate the reflection overhead introduced by the runtime architectural object, we deployed the involved components in two different kinds of JBoss AS with each running on three PCs: one is augmented with architecture listener plug_in (as an mbean), the other is not. Because of the need to exclude other interfering factors like http parsing, etc. and to calculate the computation time cost only, this experiment is done in a non-http way, and we directly invoke the interfaces of the entry component. The average performance result of the two is shown in Figure 4.9. We can see the additional cost of reflection is around 10% on average.

The time for the runtime architecture object generation and substitution (2-phrase commitment protocol) are given in Table 4.1. We can see substitution takes a longer time than the generation does. This results from the two phase

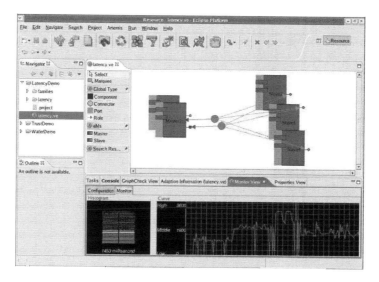

Figure 4.8: After adaptation

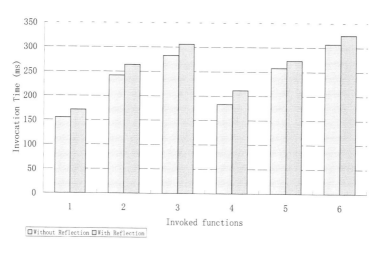

Figure 4.9: Performance comparison[†]

[†]The numbers 1,...,6 in Figure 4.9 denote functions: *ISGetIdentity, ISSearch, ISBook, IMRefresh, IMSearch,* and *IMBook,* respectively.

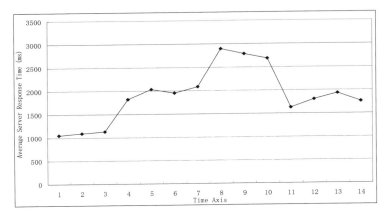

Figure 4.10: Server's average response time statistics

commitment protocol we use in the framework to ensure the atom properties of the architectural object.

Figure 4.10 shows the average response time to the http clients during dynamic evolution. At the time axis, each number denotes the time interval at around 2 seconds. The first 3 intervals represent 200 concurrent http requests, then the next 4 intervals represent 400 concurrent http requests, and the last 7 intervals represent 600 concurrent http requests. With the increase of the concurrent http requests, the server's response time increases too. As from the 8th interval, the response time exceeds the threshold set by the rule, and it triggers the event of adding a new back-up server. At the 11th interval, the response time has been reduced to the value below the threshold. The applications architecture is evolved from 1 master, 3 slaves to 2 master, 3 slaves dynamically. This adaptation process is also displayed from the assembler's perspective as illustrated in Figure 4.7 and Figure 4.8.

Based on the above experiment results, we can conclude that: 1) compared to direct invocation, the additional cost caused by the reflection is marginal with the value around 10%. This cost is acceptable in most cases; 2) the substitution cost of architectural object is higher than that of the generation, but still it is not expensive. In our case it is around 300 milliseconds; 3) the dynamic adaptation of the architecture object which redirects the invocation direction is effective to improve the performance according to the results.

4.4   Case Study

The development of information technology provides an efficient and low-cost way to everyday life. More and more people and organizations leverage the Internet to manage their work and other activities. Online systems become an important application domain of software technology. Meanwhile, the diversity of connected computing platforms and the uncertainty of the environment pose high requirements for the adaptability of these systems. In certain areas, such as water management, they usually involve broad scales, various kinds of platforms, complex context information, etc. In order to keep users' sufficient satisfaction, these contextual concerns have to be considered during the system design. Basically, there are the following requirements:

- Uniform context models. A context model with uniform semantics is the basis for efficient adaptation. The model should address the platform properties, the internal structures, and the users' profiles, etc.

- Dynamic service composition. Applications can be developed based on the composition of the underlying atom service components. These components can be dynamically incorporated and managed at runtime.

- Adaptation logic expression mechanism. The adaptation can occur in both the functional aspects and the non-functional aspects. The adaptation logic should be able to express these requirements. Furthermore, because of the complexity of the environment, in many cases, the dynamic loading of unplanned adaptation logic is much needed to improve the systems' availability.

In this section, we will use a water management application as a concrete scenario to illustrate the adaptation requirements and the solutions provided by our framework.

4.4.1 Scenario statement

In China, some large rivers usually flow through several administrative regions. Monitoring the river conditions is an important task for water conservancy management. Generally, there are some static observation spots, as well as, some mobile spots. Each spot can publish the water and rain information of the particular area. By integrating these separate services, a comprehensive water information system covering the whole flow area can be developed. For ease of use, the composed framework provides a web interface. Users can access the system through web browsers. By clicking on the particular spots in the electronic map, the related information, such as the flow conditions, the rain conditions, and the water quality conditions, can be displayed accordingly. Because of the close relationship between water resource management and the weather conditions, the application also needs to incorporate the weather forecast services.

When flood season comes, some dangerous events might happen within the river's flow area, such as, bank-burst and piping outflow. As these dangerous places have no signs beforehand, it is possible that there are no static monitoring utilities located. In these urgent cases, administrative organizations mostly will send out some automobiles with corresponding devices to collect data in these places. In order to facilitate the scheduling and arrangement work of the whole flow area, these data can be published as a service and integrated into the application to provide support for decision-making. As the new service is dynamically integrated into the system, there is a need for self-adaptation. Usually, the static monitor sites publish services through a private network. But for the mobile sites, there are possibly no such infrastructure deployed there. It has to use the public network to send the data. For security issues, it is reasonable to automatically use an encrypted channel to interact. Besides, numerous access devices exist, such as smart phones, notebooks, and desk-tops, etc. These platforms have different physical properties. The interfaces should be able to adapt themselves according to the differences.

4.4.2 Adaptation requirements

Water resource management system is an integrated application running on the Internet. It consists of several distributed services: 1) information display services; 2) water and rain information publishing and retrieving services; 3) weather information publishing and retrieving services. These atom services are usually developed and provided by different vendors. For example, the information display services are usually developed by the service integrator, and the water and rain information publishing services are usually provided by the corresponding monitoring site. The weather information services are usually provided by the local weather agencies.

Using distributed component-based techniques to implement these services is an adequate choice. However, different services might use different component techniques, such as, EJB servlet. How to integrate and coordinate these services which might use heterogeneous underlying techniques is the first adaptation requirement.

In urgent cases, such as bank-burst or pipe-overflow, some mobile monitor autos will be sent out to the places where the events happen. The autos will collect data locally for corresponding officers and experts. As there is no foreknowledge as to where these events would happen, these mobile monitors must be integrated dynamically after the system is running. When the dangerous situations have been processed, these mobile monitors will finish their task and go back. Thus, in the application, these component services will be removed dynamically in the running application. The dynamic manipulation of corresponding components from the running application is the second adaptation requirement.

Because of the lack of the foreknowledge as to which places will have dangerous water conditions, it is possible that these potential critical places have no private network deployed. In this case, it has to use the public network instead. As security issues are introduced by the public network, an encrypted channel is much needed to prevent the data from being tampered. Therefore, the support of the dynamic addition of the encrypted interaction mode is the third adaptation requirement.

As weather information is closely related to water management, weather forecast services need to be integrated into the system. There are several means to acquire weather information. For example, there are YAHOO weather services and local weather services. In general, the local weather services are more accurate and timely. Thus, the water management system would integrate the local weather services as the first priority. If this kind of service is unavailable locally, the system would automatically adapt to integrate the general weather forecast services. Moreover, the protocols between the local and the general weather services are not necessarily the same. To automatically select the adequate weather forecast services and to adapt to the corresponding protocols at runtime are the fourth adaptation requirement.

In urgent cases, the related workers and experts will communicate very frequently. They will use various devices to get the information and communicate. For example, they may use traditional notebooks, as well as, smart phones to get the most recent information about certain flow areas of a flooding river. These computing devices have different physical properties. The display unit of the application needs to adapt automatically according to the difference. This is the fifth adaptation requirement.

To summarize, the five adaptation requirements in this scenario are as follows:

1. The requirement to integrate the distributed and heterogeneous service components dynamically;

2. The requirement to manipulate (addition, removal, etc.) the service components at runtime;

3. The requirement to add the encrypted communication channel to the mobile monitor stations automatically;

4. The requirement to adaptively choose appropriate weather forecast services at runtime;

5. The requirement to adjust the display unit of the applications according to the different physical properties of the clients' devices.

4.4.3 Solutions based on MAC-ng

As stated in previous chapters, we use software architectural styles to describe the structural relations between the components and the connectors to specify the interactions between components.

For this specific scenario, the master/slave style is the most appropriate to

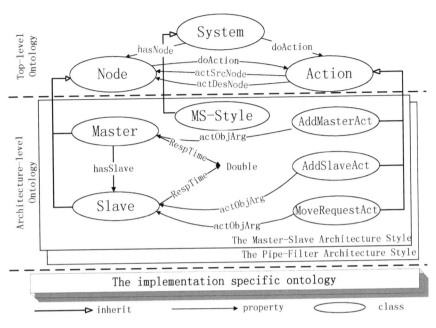

Figure 4.11: Master/slave architectural style ontology

construct the application. The central display component functions as the master node, receiving client requests and dispatching the task to separate nodes. The water and rain information publishing components and the weather forecast components function as slave nodes. Therefore, the master node has a *provide* type interface for users' looking-up and several *request* type interfaces for interaction with the slave nodes. Slave nodes have two classes, *i.e.*, water information slave and weather forecast slave. Water information components have a *provide* type interface for the related information looking-up, and the weather forecast components also have a *provide* type interface for the weather information retrieval. Each static monitor site has a corresponding water and rain information service component for the selective areas. Central and local weather bureaus have their separate weather forecast services. Figure 4.11 displays the master/slave architectural ontology model pictorially. MAC-ng provides direct support for graphical composition of the available services. It uses a graph grammar directed editor to ensure the composition process in a disciplined way. Figure 4.12 illustrates the global architecture of the application drawn in MAC-ng. The rectangles represent the functional components while the circles represent connectors. The functional components are published as services, identified by their business names and can be looked up in our framework. The interactions between the components are captured by the notion of connectors. In this example, there are two types of connectors. The first one is to specify the interaction between the master component

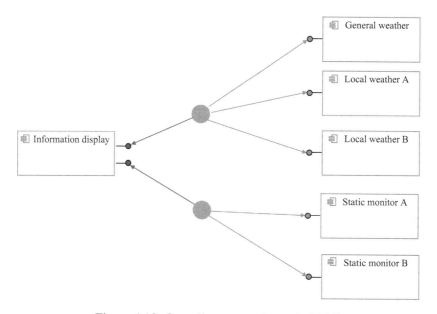

Figure 4.12: Overall structure drawn in MAC

and the weather information service components. It is responsible for the dynamic selection of the adequate weather components at runtime, as well as, the normal communication between the master component and the selected weather information component. The second one specifies to the relations between the master component and the static water information monitor components. This connector mainly takes charge of integrating the information from specific water information monitor components.

As stated previously, the components are usually developed by different vendors and possibly heterogeneous. By exposing their interfaces and the protocols, this heterogeneity can be coped with by the introduction of connectors. The connectors can consist of a set of interceptors. These interceptors are programmable and can be dynamically loaded during runtime by the language support. In this example, the first connector is composed by the *adaptive selector* which selects the adequate weather services according to the context and several *remote invocation* connectors which interact with the selected weather services. Figure 4.13 illustrates its composition.

The second connector functions like a multiplexer, it receives the invocation command from the master component and dispatches it to multiple slaves. After receiving the results from these slave components, it is also responsible for integrating them into a single message and sending back to the master. Figure 4.14 displays the structure of the second connector. MAC-ng provides templates for constructing these interceptors. Users just need to inherit the corresponding par-

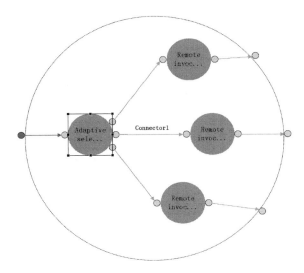

Figure 4.13: The first connector composition

ent classes and implement specific functions. These interceptors can be loaded and invoked automatically by the PCM module in our framework.

Listing 4.2: Smartphone modeling illustration

```
<owl:Class rdf:ID="SmartPhone">
  <rdfs:subClassOf>
    <owl:Class rdf:ID="Phone"/>
  </rdfs:subClassOf>
</owl:Class>
<owl:FunctionalProperty rdf:ID="hasBrowser">
  <rdfs:domain rdf:resource="#SmartPhone"/>
  <rdfs:range rdf:resource="#Browser"/>
  <rdf:type rdf:resource="http://www.w3.org/2002/07/owl#ObjectProperty"/>
</owl:FunctionalProperty>
<owl:FunctionalProperty rdf:ID="hasScreenSize">
  <rdfs:domain rdf:resource="#Phone"/>
  <rdfs:range rdf:resource="#ScreenSize"/>
  <rdf:type rdf:resource="http://www.w3.org/2002/07/owl#ObjectProperty"/>
</owl:FunctionalProperty>
<owl:AnnotationProperty rdf:ID="phoneName">
  <rdf:type rdf:resource="http://www.w3.org/2002/07/owl#DatatypeProperty"/>
  <rdfs:range rdf:resource="http://www.w3.org/2001/XMLSchema#string"/>
</owl:AnnotationProperty>
<owl:Class rdf:about="#ScreenSize">
  <owl:equivalentClass>
    <owl:Class>
      <owl:unionOf rdf:parseType="Collection">
        <owl:Class rdf:about="#Large"/>
        <owl:Class rdf:about="#Medium"/>
        <owl:Class rdf:about="#Small"/>
      </owl:unionOf>
    </owl:Class>
  </owl:equivalentClass>
</owl:Class>
...
```

The composition view of the connectors provides a flexible mechanism for

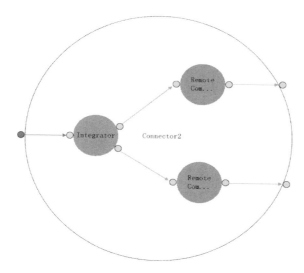

Figure 4.14: The second connector composition

the dynamic adaptation of involved components. In this example, to dynamically add or remove a weather forecast service component, users just need to reload the updated interceptors. The updated information will be sent to corresponding sites by the architectural agent.

There is numerous communication equipment with different physical properties. For example, the notebook and smart phone have different screen sizes and computing abilities. Phones are much easier to carry compared to personal computers. To provide user-friendly interfaces requires the context-awareness. We use ontology to model the context. The following fragment Listing 4.2 gives a partial definition template of a smart phone with a certain screen size and a browser type. In this model, the fragment defines that a cell phone has a screen, and a smart phone is a subclass of a cell phone which has a certain type of browser. In real applications, the instance of the smart phone ontology will specify concrete screen size and browser type. Similarly, other devices, such as smart phones and notebooks have different physical properties. The user interface of the web application will adjust its resolution rate according to the screen size automatically.

The integrator assembles the components, connectors, and the interceptors according to a certain architectural style and binds corresponding information to the architecture elements. Our framework will automatically check the result against the style constraints and transform them into the architecture configuration and the connector description files if it is a valid architectural object. The architectural information is embedded within the architectural agent and will be sent to the component container of the involving service component. After

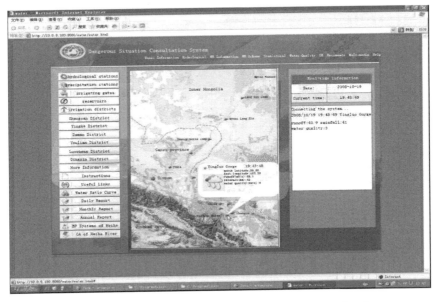

Figure 4.15: The running view of the composed application

deploying these components, the final application can start running. The web interface is illustrated in Figure 4.15.

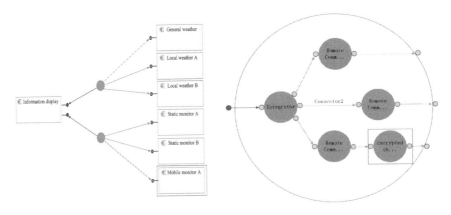

Figure 4.16: Mobile monitor addition   Figure 4.17: Interceptor addition

In urgent cases, water conservancy would send out some mobile observation devices to the scene. Reflected on the application, a new monitor service is added dynamically. As stated previously, this adaptation is accompanied by the addition of an encrypted channel in order to protect the communication from being tam-

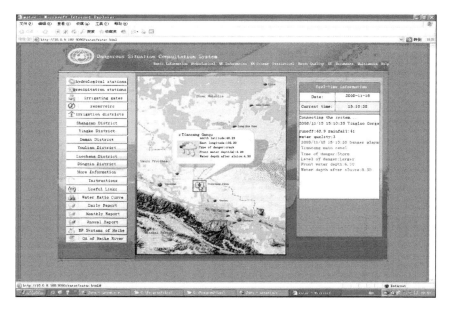

Figure 4.18: The web interface after adaptation

pered with. These adapted architectural views are described by Figure 4.16 and Figure 4.17. The newly added parts are marked up by a red rectangle separately in the figures.

The updated architecture description and the connector specification files together with the serialized class files of the interceptor will be sent to the involved sites. The server-side listener will adjust the architectural agent structure accordingly. In Figure 4.18, the adapted application interface displays the water and rain information reported from the mobile monitors. On the map, the star denotes the place where those dangerous events happen. The texts within the red rectangles denote the real-time water and rain conditions sent from the scene by the newly added mobile component.

4.5 Summary

In this chapter, we have presented MAC-ng: a platform that supports self-adaptation of distributed component-based systems. We try to provide a disciplined way to dynamic evolution instead of the *ad-hoc* ways that are commonly used. By leveraging the ontology associated with formal reasoning ability to model the context and the runtime architecture object to capture the invocation between involved components, the changes posed by the environment and triggered by the formal reasoning, can be reflected on the update of the automatically generated architecture objects. The evolved architecture object reinterprets the invocation relations. The experiment results validate our approach. The conflict detection and the grammar-directed editor can help to ensure the consistency control during adaptation.

Moreover, we also introduced a concrete scenario application in water conservancy management domain. First, we list some adaptation requirements of this scenario. Based on these requirements, we proceed to show the development support from our framework: MAC-ng. The scenario application incorporates the concerns of context modeling, architectural agent generation, and adaptive multi-mode interaction. The adaptation result illustrates the feasibility of several proposed adaptation mechanisms.

Different from traditional service-centric systems, the water conservancy applications have higher demands on continual availability. To cope with unplanned changes, the supporting framework must provide more flexibility. We use a programmable connector based way to handle this requirement. Combined with context modeling techniques and architectural style constraints, the framework offers a comprehensive solution for adaptation in such domains.

ADAPTIVE MIGRATION, SERVICE DISCOVERY, AND INTERACTION

Chapter 5

Adaptive Component Migration

CONTENTS

---

*Part of this chapter is modified and updated from [198], with permission from IEEE.

With the development of computing technology, its focus has shifted from the *machine-centric* paradigm to the *user-centric* in an open environment. As users are not constrained to interact with desktop computers in their workplace, there are emerging requirements that they should communicate with computing resources anywhere and at anytime. This is also the vision of pervasive computing: Users have personalized preferences and personal operation habits. In the new environment where they arrive, it may take some time for them to be familiar with the contextual conditions. Therefore, application mobility is an efficient way to mask uneven conditioning and reduce users' distraction in an open environment. However, since mobility brings more dynamism and uncertainty, it raises new challenges in developing pervasive applications, including underlying application models, adaptive resource rebinding mechanisms, synchronization, and fault tolerance techniques, etc. In terms of software architecture views, application mobility is a kind of reconfiguration of software's deployment views, while the logic view usually stays the same. Inspired by software agent's inherent ability of autonomy and mobility, we investigate its potential usage in the aspects of application mobility. The architecture topology information is embedded in the software agents. As the component interactions are intercepted by the agent and the agent can update its information according to environmental changes and the architecture information in the case of mobility, the adaptive component migration can be achieved. Four salient features are emphasized in our approach:

1. Reduced mobility overhead. Flexible bindings of application components can avoid migrating the whole application;

2. Simplified mobility management. From software architecture perspective, during migration, the logical view usually stays the same, although the physical view or deployment view changes. Therefore, we employ a specific entity, *i.e.*, mobile architectural agent, to take over the responsibility of mobility and synchronization, so user intervention is reduced;

3. Enhanced customizability and adaptability. Context information can be updated dynamically, and ontology-based reasoning ability embedded in the management layer can direct the application to adapt to the changes accordingly;

4. Model level specification. We use attributed graph grammar to constitute the underlying mobility semantics. In this way, some properties can be checked formally, such as whether the given deployment conforms to the style, or whether the migration strategy violates the conditional constraints.

The rest of this chapter is organized as follows. In Section 5.1, some background information is presented. In Section 5.2, we discuss architectural requirements. Based on these requirements and previous analysis, we present the design of our architectural support in Section 5.3. The attributed graph grammar based modeling and analysis is given in Section 5.4. Implementation of the applications in our proposed framework and their performance evaluations are described in Section 5.5. In Section 5.6, we review the related work. Finally, Section 5.7 summarizes this chapter.

5.1 Background

Rapid progress has been made in the recent years towards the integration of cyber space and the physical world. Various kinds of smart devices and sensors have appeared in the market place. Meanwhile, computers with higher processing ability and lower prices are diversified into common consumer electronics connected by many kinds of networks. In this computation-intensive environment, how to coordinate various kinds of smart devices and make them serve people in a more natural and satisfying manner becomes one of the main research concerns in both the academia and the industry community. Users have specific operation habits and preferences, and when they move from one place to another, it may cause some inconvenience in the new environment. For example, if a person is left-handed, he will certainly feel uneasy to work in a right-handed application environment where he moves. We observe that what the user actually interacts with on these physical devices is essentially logical software applications. If the application can migrate with the user or be customized according to his preferences, and adapt to new environment proactively, it will become personalized and thus can naturally reduce users' distraction.

However, making the application mobile, personalized, and adaptable faces several challenges. The two most fundamental problems are *when* and *how* to migrate and adapt the application. Different devices usually have different properties, such as screen size, resolution ratio, and computation capability. Thus, one application running well on one device does not necessarily mean that it would work well without any adaptation on another device. Moreover, there are multiple migration modes. For example, cut-paste like migration and copy-paste like migration. By cut-paste like application mobility, we mean that, applications (or parts of applications) save the states and migrate to the destination. By copy-paste like application mobility, we mean the application clones first and migrates. We use a metaphor to express this, as it is very like the everyday text editing operation. In the latter case, some synchronization channels need to be established between or among the involved applications. These migration strategies depend on the specific contextual requirements. As described in previous chapters, modern software applications are usually constructed through the composition of multiple components, *e.g.*, user interface component, business logic component, and resource binding component, etc. The logical architecture view of the application after migration is usually the same as that before migration, although some components may reside in new containers. We can address the problem from the architecture perspective and investigate the underlying management mechanisms of mobility, application architecture, and resource matching, etc. In addition, to capture users' movement and intention also requires the attention on context modeling and reasoning capability. The issues thus stretch from the application layer to the context layer, while the current software system offers limited support for mobility and context management. Putting all these concerns in the application layer would be too much for application developers.

The above observations motivate us to approach the issue from the architecture perspective and to offer an architecture-level support for application mobility. Inspired by the coincidence of software agent's inherent features and pervasive environment's requirements, we investigate and exploit the agents' potential usage in adaptive application mobility to meet users' needs. Software agents generally contain two complementary semantics. The first is on mobility, and the second is on autonomy. Mobile agents (MAs) are programs that can migrate in a network at times and to places of their own choosing [114]. Various kinds of agent-based solutions have been proposed and proved to be feasible and efficient in a considerable amount of applications, ranging from software engineering to knowledge engineering.

Based on the preliminary work on agent-enabled application mobility [189], we extend the research into the development of the underlying application model, synchronized mechanisms and adaptation techniques. Moreover, we embed the architecture information into the mobile agent, which makes it a mobile architectural object (we still call it mobile agent for short in the following). The previous work has proved the feasibility of agent enabled application mobility, but it does not further investigate the component-level migration, the clone-dispatch application mobility, and resource description and reasoning mechanisms. Compared with other works on application mobility in a pervasive environment [190], our approach highlights the following characteristics: 1) agent-enabled loosely coupled application architecture and flexible resource binding mechanisms support light-weight transmission; 2) agent-based coordination mechanism supports not only the follow-me kind of mobility, but also the clone-dispatch kind; 3) embedded logic-based reasoning utilities support adaptive migration behaviors. Besides the above features, employing agents can also leverage the existing methodology and architecture, thus getting the advantage of simpler persistence and mobility management as well as stronger resilience capability [114]; 4) in the meta level, an attributed graph grammar based model provides the underlying semantics for the migration. In this way, not only can the migration be formally specified, but also some properties can be checked.

5.2   Architectural Requirements

In this section, we identify some key requirements that should be addressed in the architecture design for adaptive component level mobility in an open environment.

5.2.1   *Application model*

Component level migration is more desirable than migrating the whole application in terms of cost. This requires the applications to support flexible component binding and composition. Usually, an open environment, for example in pervasive computing, offers various kinds of network connections. By utilizing this infrastructure, applications can be designed as a collection of reusable distributed objects. The requirements of the application model can be summarized as follows: 1) applications should be decomposed into separate parts, such as logics, presentations, resources, and data, so as to support loose coupling; 2) to coordinate these components, synchronization mechanisms need to be provided; 3) before and after migration, application states should be consistent and continual, so a state manager component should be provided; 4) multiple kinds of devices and network conditions exist, so the adaptation mechanisms are also required. Application models that take the above requirements into consideration can significantly ease this process to achieve environment-adapted and user-customized application level mobility.

5.2.2   *Mobility management*

What distinguishes application mobility from other kinds of mobility (such as data mobility, etc.) is that application is a proactive and executable entity. After migrating to the destination, it can continue its execution in the new environment. Basically, there are three dimensions to consider. First, which components should be migrated? Second, where is the destination, in the same virtual space or across the space? Third, what kind of migration style is needed, cut-paste like or copy-paste like?

Figure 5.1 gives an illustration of mobility classification. In the dimension of mobility modes, there are two categories, one is following the user's location and the other is cloning itself and dispatching the cloned one to the destination. Application-level clone and dispatch modes are intuitively similar to copy-paste and cut-paste operations respectively, but involve many more concerns. In some cases, we need cut-paste like mobility. For example, a person is listening to a piece of music, but he has to go to other places for some reason and he does not want the music to be interrupted. Now, the best way is, the music player can stop when he moves out (cut) and continue when he enters the new place (paste). While in other cases, we need copy-paste like mobility. For example, in

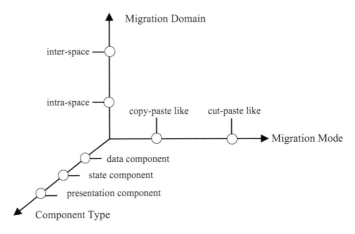

Figure 5.1: Mobility classification

conference scenarios, we often face the following embarrassment, one or several members cannot come due to various kinds of reasons and, in this way, the meeting applications might clone themselves (copy) and move the copy to the destination (paste). The application would start automatically and synchronize with the source application at the destination. Along with the voice transmission, a remote meeting would be made possible.

In the dimension of mobility domain in Figure 5.1, due to the current technology limits of coverage, generally one smart space only covers a specific area. Migration across the space boundary requires additional gateway support. Also, applications should be aware of which parts of the components can be migrated, the data, the presentation, the logics, or other components. The mobility management design should take these into consideration.

5.2.3 Resource binding and service customization

As discussed in Section 1, after the migration of the applications' components, for various reasons, the original resource bindings may be lost. For example, if the network is busy and the destination machine has the required resources, then the local resource can be used without the need to transfer resources from the remote source host. This requires a resource rebinding mechanism. As different hosts often have the same resources but with different names, simple syntax-based matching sets many unnecessary constraints, while semantics-based resource matching is much more preferable.

Service customization has two categories. The first is for different devices, while the second is for different users. This requires explicit specifications for these two cases and an introspective ability of applications to adapt to different scenarios.

5.2.4 *Context awareness*

To capture the users' mobility or intention, the application should be aware of the users' current context, which involves the information that can be used to characterize the situation of an entity relevant to the interaction between a user and an application [2]. Since application's mobility and customizability are closely connected with the users' locations and personal preferences, in system design, this kind of context should be specifically paid attention to. Different context information often has different properties. For example, the users' location information usually changes frequently as people often move from one place to another, while the users' preferences or operational habits are generally more stable. Modeling different context information also requires taking their temporal characteristics into consideration. Usually, the underlying sensors can only collect raw data such as distance, and badge (listener) identity. To map the data to useful information such as location, user identity requires context fusion mechanisms. Besides, some context reasoning and prediction functionalities should also be provided to improve the performance.

5.3 Architectural Framework

5.3.1 *Application management*

In this section, we introduce our architectural framework to address the issues discussed above. To support highly customizable and adaptable applications, we adopt a loosely-coupled application model, which follows a distributed model-view-controller pattern. Presentation component is mainly designed for the user interaction. Business logic is embedded in the controller component. Data and other resources are managed by the model component.

5.3.1.1 *Application architecture*

Our application architecture has two layers as shown in Figure 5.2. Upper level is the application layer which consists of some application components, such as the logics, the presentations, and the resources, etc., together with some description files, such as the user profiles, the device profiles, the resource profiles, and the interface descriptions. Logic controller handles the processing of data and resources and controls the presentation components. As this level directly interacts with users, it is visible to them.

At base layer is the management layer. The main modules are coordinator, snapshot manager, persistent management, mobile architecture agent, context reasoning engine, and adaptor. The coordinator establishes the synchronization link between different presentations and interacts with snapshot management and mobile agent. Basically, different presentations register themselves to the coordi-

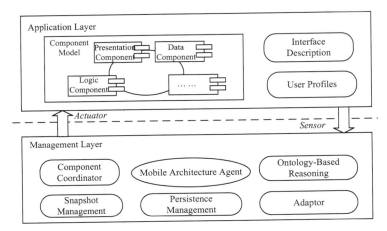

Figure 5.2: Application model

nator. When the states change, these presentations can get notified automatically. In this way, we can get not only a loosely-coupled architectural model but also simplified consistence control and higher component reusability. The snapshot management is responsible for the interaction with the persistence management module to manage the states of running applications, while the mobile architecture agent is for application wrapping and migration to the destination. Due to the high dynamism and variety in the environment, the adaptor comes to bridge the mismatch. As this layer mainly deals with underlying supporting work, it is transparent to end users.

The sensor layer will collect data from these physically or logically deployed sensors detecting users' mobility, network connectivity, latency, etc. The information will be fed into the management layer. In many cases, the raw data collected cannot be used directly at the upper level, due to the variety and frequent inaccuracy of these data sources. In the management layer, first, a classifier component will store the data into different databases according to their temporal characteristics. Mobile architecture agent (MA) is the key to connect the management layer and the application layer. The context reasoning module is responsible for reasoning and decision-making according to the data received from the context layer. Mobile agent (MA) is responsible for the wrap of application components. They communicate through message passing. When sensors detect the user's movement or indication to move an application to a remote host (cut-paste kind or copy paste kind), it first notifies the MA to prepare to migrate and record the application state. After arriving at the destination, MA retrieves complied resources and application information (maybe the information is owl-coded as it can be matched in a semantic way) from the registry center. Then according to the result and the application-specific rules, the coordinator component decides whether to transfer the states only or the interface only or other possible component combinations in application layer. Mobile agents will take over the next transmission and synchronization work according to the application-specific requirements and interact with the snapshot management module and the persistence management module.

5.3.1.2 Dynamic interaction

Applications first register themselves to the application and resource registry centers with their interface descriptions and other parameters such as specific device requirements and user preferences, etc., in a WSDL-like format. When the mobile agent gets the message of migration, it first parses the scripts. If the migration is follow-me like, it contacts the registry centers first to check whether the destination has the corresponding components and resources. Then it suspends the current execution of application, collects and wraps the snapshots together with corresponding components, and migrates to the destination hosts and resumes application execution there. If the migration is clone-dispatch like, it also looks up in the registry center first to find whether the destination host has required resources and components.

After migration, the application needs to be adapted in the new environment, and the mobile agent will contact the adaptor to conduct necessary adaptations according to some customizable parameters to adjust certain sizes, resolutions, etc.

5.3.1.3 Coordination management

In our model, the mobile architecture agent functions like a coordinator and weaves the applications and the context management. They collaborate together and interact closely with both the application layer and the management layer.

The coordination manager mainly has the communication and coordination utilities and serves as a rule manager for contextual information. The context observer continually monitors and broadcasts the context information. Not all of this information is useful. Some are duplicates and some are irrelevant. The management module will filter and find their interested subjects and interpret them accordingly. For example, when the context observer finds the user's location being changed and announces this event, this information will be captured and interpreted as the user will leave the room and inform the coordinator. The coordinator will subsequently call the snapshot manager to record the current application states if necessary, and then suspend the application. When the user's new location is announced, the coordinator module will first check application related profiles including resources, preferences, and device properties. Then it will contact the registry center to retrieve destination environment information, such as, whether the devices are compatible, if the application components exist there, and whether the network situation allows the local data to be copied. Based on the above considerations and user defined rules, the coordinator will decide

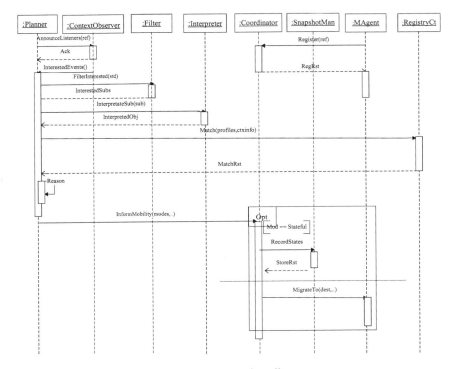

Figure 5.3: Interaction diagram

whether and what parts of the application will be shipped to the new environment through a message to the mobile agent manager.

Mobile agent will wrap the corresponding components, check out from the current site, check in at the destination, inform the coordinator to establish the synchronization link if necessary, and resume the execution. The interaction is pictorially described as a sequence diagram in Figure 5.4. In this way, the mobile agent is not bound to a specific component of applications; instead, it can wrap up any serializable part and migrate to the destination.

5.3.2 Resource description and agent reasoning mechanism

In pervasive environments, various kinds of resources with different properties exist. Some are transferable, but others are not; some can be easily substituted, but others cannot. For example, a printer is not transferable but can be substituted, while the database is neither transferable nor easily substituted, and a smart device is transferable but not easily to be substituted as the users' profiles and preferred software are installed. In order to share and utilize these resources, a representation framework is needed. We use ontology to model the resources and their inter-relations, as not only does it support resource matching semantically, but also facilitates the reasoning process.

Listing 5.1: OWL description illustration

```
<owl:Class rdf:ID = "hpLaserJet">
 <rdfs:comment>hp color printer</rdfs:comment>
  <rdfs:subClassOf rdf:resource="#Printer;Substitutable;UnTransferable"/>
   <owl:ObjectProperty rdf:ID="locatedIn"><rdfs:range rdf:resource="#Office821"/>
   <rdfs:type rdf:resource="TransitiveProperty"/>
   </owl:ObjectProperty>
  ...
</owl:Class>
```

In the domain of knowledge-based systems, ontology means a specification of a representational vocabulary for a shared domain of discourse – definitions of classes, relations, functions, and other objects, as in software literature, what "exists" is exactly what can be represented [85]. To support ontology, one description mechanism must be selected. We choose Web Ontology Language (OWL) for its generality. OWL [7] is a semantic markup language for publishing and sharing ontology proposed by W3C's Web Ontology Working Group. It is developed as a vocabulary extension of Resource Description Framework (RDF). OWL follows the XML syntax and has the advantage of platform-independence. For example, we can define a specific printer as in List 5.1.

By abstracting and specifying some key properties in the OWL format, we can check the resource compatibility semantically and customize the application according to the checking results and other context information. First, the coordinator component will retrieve the resources available in the destination host from the registry center in the standard OWL Query Language (OWL-QL) and then carry out the compatibility checking using predefined rules which can be encoded in an RDF format as the following script shows. The example script in Listing 5.2 means that the predicate 'locatedIn' is a transitive property; if the resources in the source and destination are both the 'printer' type, then they are compatible; if the resources in the source and the destination are compatible and the network condition is good (response time is less than 1000 ms), then a move command will be issued which then is to be transformed into a concrete action.

When MA gets to the destination and resumes the application there, it will also check with the coordinator to make some adjustments according to the environment configurations.

Listing 5.2: RDF rule illustration

```
[Rule1: (?p imcl:locatedIn ?q), (?q imcl:locatedIn ?t) ->
(?p imcl:locatedIn ?t)]
[Rule2: (?ptr imcl:printerObj 'printer'), (?srcRsc rdf:type ?ptr),
(?destRsc imcl:printerObj ?ptr) -> (?srcRsc imcl:compatible ?destRsc)]
[Rule3: (?addr1 imcl:address ?value1), (?addr2 imcl:address ?value2),
(?srcRsc imcl:compatible ?destRsc), (?n imcl:responseTime ?t),
lessThan(?t, '1000'^^xsd:double) -> (?action imcl:actName "move"),
(?action imcl:srcAddress ?add1),(?action imcl:destAddress ?add2)]
```

5.4 Modeling by Attributed Graphs

For the component-level migration, the interaction between the application components and the physical devices is concerned. A combination of physical view and process view will better reflect this dynamic interaction. Graphical notations, such as an UML deployment diagram, can give an intuitive description and provide a basic tool to analyze some properties, but they lack a formal background. In this chapter we will use the attributed type graph as the underlying formal semantic foundation to describe the process and use the graph transformation to model the migration process. In this way, some properties or constraints can be analyzed and verified formally.

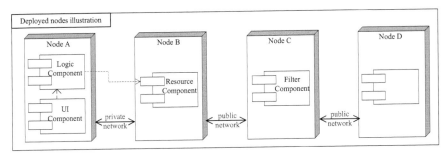

Figure 5.4: Application deployment diagram with UML

Figure 5.4 gives a description of an example deployment diagram. An application is composed of three main components, *i.e.*, *UI*, *Logic*, and *Resource*. The *UI* and *Logic* are located in Container A. The *Resource* component is located in Container B. A separate component *Filter* is located in Container D. A and B are within a private network, while B, C and C, D are connected by public networks. For simplicity, some additional properties are omitted in the graph, such as the interfaces and protocols, etc. The diagram displays a current application deployment state. As discussed previously, the UML diagram provides an intuitive description but lacks a formal treatment of underlying semantics. Besides, it is difficult to describe the facet of component migration style, *i.e.*, the migration possibilities. Therefore, it is difficult to check whether or not the migration violates any constraints. In the following, we describe the technique put forward in [198] based on graph grammar theory which can mitigate the aforementioned difficulty.

Definition 5.1 Application Deployment Graph (ADG). Given a label alphabet L_T with terminal symbols and a signature $Sig_{adg} = (S, OP)$, ADG=$(V, E, s, t, lv, le, Alg_{sig_{adg}}, av, ae)$, where

- V, E are vertices and edges respectively. Corresponding to our context and application model, we define two kinds of vertices, *i.e.*, component and container. Component can be further classified into several types, such as, UI, Logic, Resource, Data, State. Edges have three kinds, *i.e.*, logical connection (component invocation), physical connection (network connection, public, or private), and residence connection.

- s and t are source and target functions from edge to vertex, *i.e.*, $s : E \to V$ and $t : E \to V$; $lv : V \to L_T$ and $le : E \to L_T$ are labeling functions for vertices and edges respectively.

- Alg_{adg} is a sig-algebra.

- $av : V \to \upsilon(AG)$ *and* $ae : E \to \upsilon(AG)$ are the vertex and edge attributing functions. The attributes include the application-specific information, such as component name, node location, resource type, and dependency relations, etc.

Lemma 5.1
For every application deployment diagram, there is a corresponding ADG.

Proof 5.1   Constructive approach:

For every UML node or component, their names constitute the set V, and their edges constitute the set E. The labels form the set L, lv maps the component and node to their corresponding labels. le maps the edges to their labels. The properties of the nodes and edges constitute the attribute set of respective entities. The data types and operations form the signature. av and ae are attributing functions which map the vertices and edges to their attribute sets.

Figure 5.5 gives the ADG representation of the deployment diagram in Figure 5.4. The application related information are mainly embedded in the attribution part. For simplicity, we do not specify attributes here in the ADG.

Based on the definitions of ADG, we can define the application deployment style graph.

Definition 5.2   Application Deployment Style Graph (ADSG). Given a label alphabet L, and a signature $Sig_{adsg} = (S, OP)$, then ADSG= $(V, E, s, t, lv, le, Z, av, ae)$, where Z is the final Sig_{adsg}-algebra. The elements: $V, E, s, t, lv, le, av, ae$ are similar to those in ADG except that Z is the operation over the sets of the sorts as it is the final algebra.

A Sig-algebra is an S-indexed family (A_s) of carrier sets together with an op-indexed family of mappings $(op^A)_{op \in OP}$ such that $op^A : A_{s_1} \times \cdots \times A_{s_n} \to A_s$ if $op \in OP_{s_1 \cdots s_n, s}$.

$\upsilon : Alg(Sig) \to Set^P$ is a functor assigning to each Sig-algebra A the disjoint union of its carrier sets A_s, and to each homomorphism f the disjoint union of the total functions f_s, for all $s \in S$.

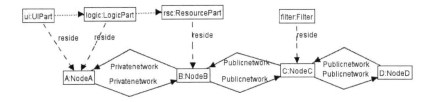

Figure 5.5: Attributed graph representation of Figure 5.4

ADSG can be used to specify all the possible configurations of a particular application deployment style. The total graph morphism $t : ADG \rightarrow ADSG$ can be used to establish the membership relationship between ADG and ADSG. Figure 5.6 illustrates one possible style of the example application. In this style, the *Filter* component can be deployed on NodeC and NodeD, connected by a public network, which means *Filter* can migrate between the two types of Nodes. *UPPart* component can be deployed on all the nodes. *LogicPart* and *ResourcePart* components can be deployed on the private Nodes only. In this diagram, we employ the inheritance representation of attributed graphs. It has been proved that there is an equivalence between the attributed graph with inheritance and without it. For details, we refer interested readers to [51].

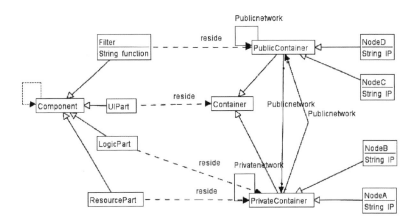

Figure 5.6: ADSG illustration

Definition 5.3 A migration transformation $MT: SubG_L \xrightarrow{r} SubG_R$ consists of a left subgraph $SubG_L$ and a right subgraph $SubG_R$, and an injective partial graph morphism r.

With the definition of MT, the two application deployment states can be connected by the *match* between the MT and the ADG, *i.e.*, a total morphism $m : SubG_L \rightarrow ADG_L$, and, therefore, the result ADG_R would be derived by $ADG_L \xrightarrow{r,m} ADG_R$.

Besides the application deployment style, we might also have some global constraints over the migration. This can be modeled by graphical constraints. A deployment constraint $c : P \rightarrow Q$ is satisfied by an ADG, written $ADG \models c$, if for all total morphisms $p : P \rightarrow G$ there is a total graph morphism $q : Q \rightarrow G$ such that $q \circ c = p$. P and Q are termed as *Premise* and *Conclusion*, respectively. In essence, Premise and Conclusion describe the coupling relations between related elements.

To be a valid component migration strategy, it has to be subject to the deployment requirements. In this part, we mainly consider the configuration validity. Concretely speaking, a *valid* migration strategy is the one that the chosen component is allowed to migrate and can migrate to the right destination; moreover, the component dependencies are to be ensured.

Theorem 5.1

A component migration strategy is valid if their host graph ADG_L and the derived graph ADG_R are both instances of the ADSG and $ADG_L \models c, ADG_R \models c$.

Proof 5.2   As specified in Lemma 1, each application deployment diagram has a corresponding graph. As validity means that the component is in the adequate location, this is specified in the *ADSG*. If the ADG_L and ADG_R are both instances of the *ADSG*, there are two total morphisms $t_1 : ADG_L \rightarrow ADSG$ and $t_2 : ADG_R \rightarrow ADSG$. As proved in [93], there is a typed attributed graph morphism $f : (ADG_L, t_1) \rightarrow (ADG_R, t_2)$, such that $t_2 \circ f = t_1$. As $ADG_L \models c$ and $ADG_R \models c$, which means the start and the derived deployment graphs comply with the application constraints, thus, the morphism f is a corresponding valid migration strategy.

Figure 5.7 illustrates a valid migration strategy. The component *UI* migrates from *NodeA* to *NodeB*. Because in the *ADSG* (Figure 5.6), *UI* can migrate among the sites. Therefore, this migration strategy is valid.

ADSG defines a set of possible configurations of application deployments. MT defines the connections between these deployments. However, there might be some additional requirements for the application deployment. For example, in Figure 5.4, the UI component can migrate from private network to public network. In this case, there must be a filter component which can provide some encryption functions for the UI part.

In the following section, we will illustrate the applications of our models in which two kinds of property checking are mainly considered, *i.e.*, whether a

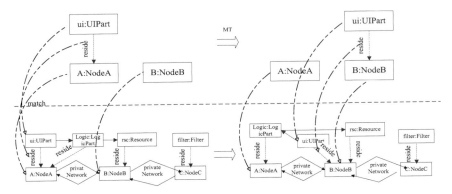

Figure 5.7: Transformation example

graph is an instance of a particular deployment style and whether the migration rule ensures to be consistent with the global constraints.

We use the attributed graph tool *AGG* to conduct the property analysis. AGG is a general development environment for algebraic graph transformation systems. It also offers validation support like graph parsing, consistency checking of graphs, as well as, conflict detection of graph transformation rules [171].

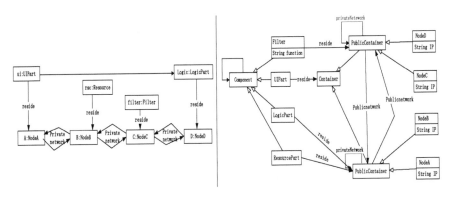

Figure 5.8: Type checking illustration

The first property checking is to find whether the current application deployment is an instance of the deployment style. The algorithm for this kind of checking is based on the total graph morphisms between the ADG and the ADSG. The checking is mainly performed in the types, attributes, algebraic specification, and equality functions.

Figure 5.8 illustrates a type checking example. In the left part of the fig-

http://www.user.tu-berlin.de/o.runge/agg/

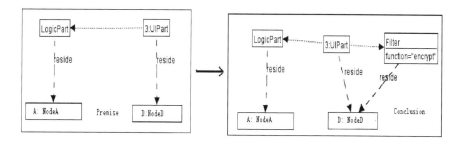

Figure 5.9: Deployment constraints illustration

ure, the *logical* component which is of *LogicPart* has migrated from *A* to *D*. Thus *source*(*reside*) = *Logic* and *target*(*reside*) = *D*. While in *ADSG*, *source*(*reside*) = *LogicPart* and *target*(*reside*) = *PrivateContainer*. There is a morphism *t* from *logic* to *logicPart*, *i.e.*, *t*(*logic*) = *LogicPart*. However, this morphism does not hold from *D* to *PrivateContainer*. Therefore, the checking fails, which means that the current deployment state is not a valid one. According to Theorem 1, the migration strategy for the start *ADG* to the *ADG'* in this example is not valid.

The other property we check is whether the migration strategy meets the deployment constraints. Figure 5.9 gives an example constraint, showing that whenever *UIPart* is located in a public node and connected to the *Logic* component, there should be a *Filter* component connected to it to provide the encryption service.

In many cases, checking whether the graph meets the constraints is time consuming, because it usually involves two steps. The first step is to transform the rule to the host deployment graph, find a match, and transform to the new deployment graph. The second step is to check whether the new deployment satisfies the constraint. We use the construction approach from Heckel and Wagner [93], to transform the consistency conditions into preconditions for individual rules. In this way, to check whether the component migration policy is compatible with the constraints, we only need to check that the host deployment complies with the application conditions.

We use the same example to illustrate this process. Figure 5.10 illustrates an invalid transformation which violates the deployment constraints. But the derived *ADG* is an instance of the *ADSG* specified in Figure 5.6. The rule specifies that *UIPart* migrates from *NodeA* to *NodeD*, but in *NodeD*, there is no filter component located. In graphical terms, we say, there is an occurrence from the *premise* to the start *ADG*, but there is no total morphism from *conclusion* to the Derived *ADG*.

To transform the constraints into the application conditions of the rule, first we need to find the minimal overlap of the *Premise* and the right side of the

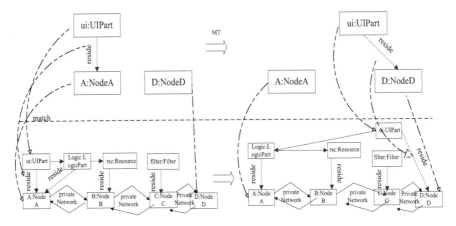

Figure 5.10: Invalid transformation illustration

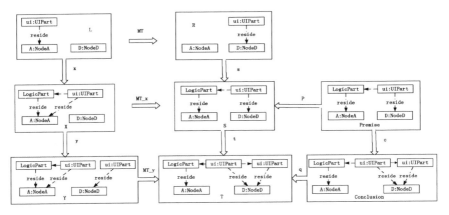

Figure 5.11: Construction of the precondition from constraints

transformation rule. Then we apply the inverse of the rule to the overlapping graph and derive a precondition. For details of the type checking, we refer readers to [58].

Figure 5.11 shows the construction of the preconditions for the *MT* rule according to the conditional constraints. The application condition in this case is: $AC(MT) = L \rightarrow X \rightarrow Y$. If for all the total injective morphism $X \xrightarrow{m} G$, with $n \circ x = m$, there is a total injective morphism $o : Y \rightarrow G$ with $o \circ y = n$, we say the total morphism $L \rightarrow G$ satisfies the constraints. And, therefore, the derived graph satisfies the constraints.

Figure 5.12 shows the rule *MT* with the application conditions constructed from the given deployment constraints. To apply the rule to the start ADG in the lower part of the figure, we find that the morphism $L \xrightarrow{x} ADG$ does not satisfy the condition c, because for the morphism $n : X \rightarrow G$ with $n \circ x = m$, there is no

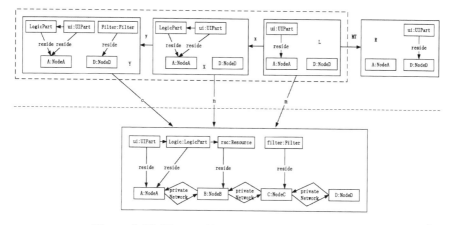

Figure 5.12: The application condition of the rule

morphism $o : Y \rightarrow ADG$ with $o \circ y = n$. So we do not need to apply the rule to the *ADG* and check the derived deployment graph to be an instance of the *ADSG* and whether it satisfies the global constraints.

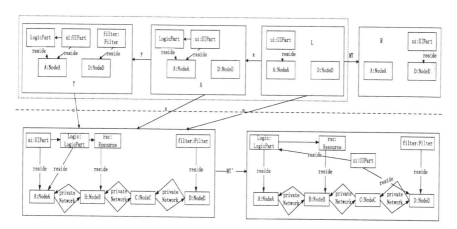

Figure 5.13: The application condition with new deployment graph

Figure 5.13 illustrates the application condition checking for a new ADG, in which now the *Filter* component is deployed in *NodeD*. We can see that now, the *ADG* satisfies the preconditions of the rule. As proved in [93], the satisfaction of the preconditions of the rule denotes that the derived graph satisfies the corresponding constraints. According to Theorem 1, we can conclude that this component migration strategy is valid.

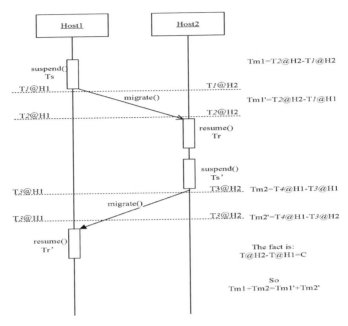

Figure 5.14: Round-trip time cost calculation illustration

5.5 Performance Analysis

In this section, we will describe the implementation of a prototype of the proposed architecture for application mobility and some sample applications built upon the framework. The prototype is implemented in Java and the agent server is JADE 4.0 [165]. We use several open source packages (in Jar file). Dozens of Cricket Sensors are deployed to collect the user's location and identity data. The prototype consists of a running kernel of context management, MA manager, coordinator module, and abstract application interfaces. Context kernel employs a publish/subscribe design pattern. When the subscribed events occur, the information will be multicast to the registered listeners. Mobile agents are implemented as specific agents inheriting JADE's Agent class. Jena [123] is used as the reasoning engine.

We have built six demo applications based on this infrastructure, namely, smart media player, follow-me editor, ubiquitous slide show, hand-held editor, hand-held music player, and follow-me instant messenger. Among these applications, we will introduce two of them as they demonstrate different kinds of application mobility. The first is the follow-me kind of music player. It can stop the music when the listener is out of the room and resume playing when the listener enters the room within the same space. In this demo, application is divided into several functional components: the codec logic, the interface, and the data files. When the context manager senses the change of user's location, it notifies

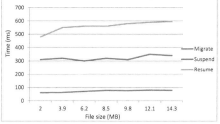

Figure 5.15: Adaptive migration cost

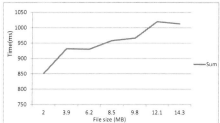

Figure 5.16: Total time cost

the coordinator. The coordinator will interpret that the user is going to leave the room and will issue a command to the coordinator to suspend the current music, as this is a stateful application, the coordinator will call the snapshot manager to record the current states. When the user enters a new place, the context manager notifies again the coordinator, which first contacts the destination hosts and checks whether the required resource or application exists or not. In this case, the resource is the music files in the playlist. If these files do not exist in the destination, they will be played remotely through URL in the original host. We use Juddi and MySQL as the backend application and resource registry center. The context management layer first check whether the application exists or not in the destination. If it exists, mobile agents just wrap the state and migrates. Otherwise, it will also carry the logics and user interface as well as the states.

To evaluate the platform performance without the loss of generality, we assume the destination host contains the application user interface but no music data or application logic. We calculate the time consumption at three phases: the suspension, the migration, and the resumption. Time consumption of suspension and resumption is easy to calculate, as both actions occur at the same place. But migration involves two places whose clocks are not synchronized. In this case, we calculate the round trip time cost. The calculation is illustrated in Figure 5.14. According to stable physical properties of crystal frequency, the difference of time values of clocks at the same time is nearly a constant value. In this way, adding up the round trip migration time cost can just eliminate the error introduced by the asynchronization of different hosts, *i.e.*,

$$T_1@H_2 - T_1@H_1 + T_4@H_1 - T_3@H_2 = T_2@H_2 - T_1@H_2 + T_4@H_1 - T_3@H_1$$

In our previous work, for specific applications, we use a static binding between mobile agents and applications. In this way, application components, including the data, the logic, and the user interfaces, all migrate with users. This will decrease the performance when the applications' size grows.

In the experiment conducted, we use different sizes of music files. The evaluation result is shown in Figure 5.15. The experiment is done on two computers

---

$T_i@H_j$ means time value at the moment of 'i', in Host (Place) j.

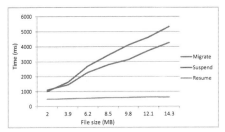

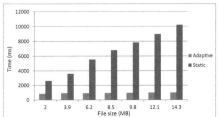

Figure 5.17: Static binding cost   Figure 5.18: Comparative time cost

both with Intel i3 2.1GHz, 2G RAM and are connected by 100Mbps Ethernet. We selected file sizes ranging from 2.0MB to 14.3MB and measured the time cost of suspension, migration, and resumption. The evaluation results imply that as the file size increases, only resumption takes more time, while suspension and migration are not much affected. Figure 5.16 displays the total time cost including suspension, migration, and resumption for different file sizes. We could observe that even when the file size increases to 14.3MB, the total time cost is a little more than 1000 milliseconds, which is acceptable in most cases.

In order to give a comparative view of the efficiency of the adaptive component migration, we have also measured the time consumption of the original design [189]. The corresponding performance evaluation and comparison are given in Figure 5.17 and Figure 5.18, respectively. From the comparison, we can observe that the adaptive migration is much more efficient than the application-level migration as file sizes increase.

The second application is to demonstrate clone-dispatch kind of migration. It needs to cross different space in our case. For example, one lecture is going to be given, but there are so many listeners that one room is not big enough to sit all of them. Some of the attendees are assigned to other meeting rooms. Traditionally, not only the audio needs to be transmitted to these rooms, additional assistants are also needed to open the slides and synchronize manually with the main room (where the speaker is).

Our demo simplifies this process by letting the agent clone the application and migrate to the other rooms and establish the synchronization links with the main room automatically. The coordinator gets the context information from the users' indication and the list of destinations. After resource retrieving and matching, it will notify MAs to migrate the components to the destination. In this case, each meeting room is equipped with a presentation application and a projector. What lacks are the slides. So MAs just need to carry the slides to the destination, collaborate with the MA manager and synchronize the slides with the speaker's presentation controls. Meantime, separate channels will broadcast the speaker's voice. So that attendees can hear the same lecture in different rooms. In our scenario, different rooms belong to different cyber domains, gateways are provided to connect them. In implementation, we import part of *Open Office Impress* as the slide show presenter and Open Office SDKs to get the controller handle. We refactored the program according to the structure model introduced previously and added the coordination components to synchronize the different presentations.

5.6   Related Work

In this section, we review some representative work on application mobility. The idea of application mobility results from the requirements of personalized and adaptive services in the open environment. Several other research projects have worked on this aspect. For example, there are Gaia [147], Aura [78], BEACH [172], one.world [84], to name a few. Comprehensive survey of research on application mobility is impractical in this chapter, thus, we select and highlight some typical examples in the literature.

Gaia is a platform which models the pervasive environment as an active space which is capable of sending the context of the user habit, can react and support customization. The application framework of Gaia uses reflection to explicitly separate the application base-level from meta-level. Applications are decomposed into five parts, namely, the model, the presentation, the adapter, the controller, and the coordinator. Computational reflection manages the complexity in the development of applications, allowing developers to concentrate on the base-level and providing mechanisms to automate meta-level configuration dynamically. The coordinator manages the component registration, application's life cycle and mobility. However placing all these responsibilities into a single static coordinator module will unavoidably increase its complexity and cause the problem of single point of failure. In addition, Gaia lacks a unified resource definition framework.

Project Aura recognizes human attention is a scarce resource in open environments and aims to offer a distraction-free framework for user mobility in the pervasive environment. To enable the distraction-free mobility, Aura needs to know the user's intents and recognize the actions. Moreover, new computing environment needs to be reconfigured based on the user's profile. Different from Gaia, applications are organized into services. User tasks become first class entities represented as coalitions of abstract services. The task manager will coordinate these services and transmit them accordingly through a file transfer system after they sense users' mobility. However, Aura did not address much about adaptation after the migration. Besides, inter-space application transmission as well as multi-application synchronization issues have not been further investigated.

BEACH is a software infrastructure providing functionality for synchronous cooperation and interaction with room-ware components. By room-ware, it refers to those room elements augmented with information technology, *i.e.*, smart rooms. It uses an event dispatching mechanism to support multiple persons using the same devices concurrently. The synchronization is realized through shared objects. When the state of these shared objects changes, the updates are triggered automatically. This is somehow similar to the update mechanism used in our system. But the emphasis of BEACH is to support synchronized multiple devices collaboration.

Similar to Aura project, one.world framework utilizes services to provide infrastructure support. To help developers in making their applications adaptable, it has a set of built-in services that serve as common building blocks, such as operations, migration, and checkpointing [84]. These services are integrated into the process manager, which is in turn responsible for checkpointing and migration, *i.e.*, to save and restore the application states before and after the application migration. Applications are composed of multiple components which communicate data in the form of tuples, and they reside in containers which is termed as *environments*. Environments manage the life cycle of inhabitant applications, including migration. But they do not support adaptive component level migration.

Previously, we also designed an agent-enabled platform supporting application level mobility, *i.e.*, MDAgent [189]. To continue along the same line, we further the investigation from the aspects of the underlying application model, the mobility management, and the separation of concerns in agents. The original framework uses a static binding between agents and applications while the current one adopts an adaptive binding mechanism, in which only parts of the applications need to be wrapped to migrate. It can help reduce the migration cost significantly, which can be shown in the performance study. Our original framework only supports follow-me kind of mobility, we have extended this to support clone-dispatch mobility. Also, the reasoning functionalities are now separated and incorporated into a specific coordination management module, while these functionalities were formerly mixed together in mobile agents. This separation of concerns also facilitates the agents design because different agents just need to concentrate on their specific roles.

There is another thread of research which can support similar user experience to application mobility, but through a quite different dimension, *i.e.*, remote access protocol based solution, such as, remote frame buffer protocol (RFB) and remote desktop protocol (RDP). Through the remote access protocol, the viewer (client) and the server could interact seamlessly. The graphical user interface will be transmitted to the viewer side and the peripheral operations, such as the input from keyboards and mouses by the clients, will be transmitted back to the server and just like that the interaction happens locally. Typical examples of RFB based solution include virtual network computing (VNC) which is open source, and the RDP based solution is Microsoft's Remote Desktop Application which is proprietary. The problem is that such remote access protocol based solution cannot support the remote computer to directly use local resources, for example, local printers.

5.7 Summary

In this chapter, we mainly introduce an adaptive component migration enabling technique in the open environment. Observing that the logical view of software architecture stays the same, while only the deployment view changes during migration, we exploited the potential use of software agents to support adaptive component-level migration. Particularly, we investigate the problem of application mobility from the aspects of the underlying application model, mobility management, collaboration of different kinds of software agents, resource matching, and service customization mechanisms. By application migration, users can interact with the environment in a more natural and comfortable way. Our experiments and experience have indicated that mobile agent technology is a promising approach to support adaptive application mobility. At the meta level, we also employ attributed graph grammar to specify the component mobility. Based on this specification, we can check some properties, such as the type conformance and the constraints violation formally. The proposed architecture has some unique features which distinguish it from other alternatives. It supports flexible and multiple kinds of application mobility. Semantics-based resource matching and reasoning mechanisms enable richer information processing.

Chapter 6

Service Discovery and Interaction Adaptation

CONTENTS

Service discovery and interaction are important, as well as challenging issues for computing in the open environment. Computing entities—either software modules or hardware devices—might migrate or exist in different interconnected cyberspace. Automatic service discovery and interaction support is necessary to improve users' satisfaction. Indeed, to date, many service discovery protocols have been proposed and a number of new ones are under development. However, an open environment usually involves applications running in heterogeneous networks and application developers must cope with the diversity of underlying network infrastructures, middleware platforms, and hardware devices. There is an urgent need of self-adaptive mechanisms to support the discovery and interaction of the services which are available in different environments.

In this chapter we present how to provide adaptive service discovery and interaction support in an open environment. We approach this problem with a flexible combination of the interceptors and propose an adaptive connector-based approach to support multi-protocol service discovery and multi-mode interaction. In software architecture, connector represents the encapsulation of coordination entities and is generally regarded as a first-class entity. In our approach, the connectors include a universal adaptor interface, protocol-specific plug-ins, and a series of interaction interceptors. An ontology-based context modeling of the protocols gives the mutual-interaction a common understanding of the environment. The protocol-specific plug-ins and the combination of the interaction interceptors can be adapted according to the requirements of the environment and the users. The interaction feature decomposition and configuration model is proposed. We also argue that a programmable coordination manner is necessary to support flexible interaction modes. This can be realized based on reflection technology. A decomposition and configuration model of interaction features can assist developers in interaction programming by analyzing and synthesizing interaction features.

The rest of the chapter is organized as follows: some background information and research motivations are given in Section 6.1. The approach including the architecture, discovery primitives, and mapping mechanisms, and interception modeling and composition, is presented in Section 6.2. The experiment is discussed in Section 6.3, followed by the review of some related work in Section 6.4. Section 6.5 concludes this chapter.

6.1 Introduction

Due to the increasing pervasiveness of computing devices in the physical world and the shift of computing orientation (*i.e.*, from machine-oriented to human-oriented), users are no longer constrained to only a single computing environment; instead, they can work across different environments and meet a diversity of software, hardware devices, and network infrastructures. Heterogeneity of the related environments brings significant problems that application developers must cope with. For instance, devices must be able to discover and share services among each other. However, manual configuration may require special skills together with a long set-up time. Therefore, adaptive service discovery and interaction becomes an important task in an open environment. However, the characteristics of services in the open environment, such as independence, diversity, and customization, may pose some difficulties to this vision. 1) Independence: the development, management, and integration of services rely on different independent organizations. Independence restricts these services to be used with little modifications. Thus, it is usually hard to satisfy specific requirements of end users. 2) Diversity: accompanying service independence, services also present complex diversity, in terms of various modeling approaches, coordination modes, and communication languages, etc. The diversity of services makes inter-operation and multi-mode interaction extremely important. 3) Customization: in an open environment, it is difficult to develop self-contained services based on a complete set of information and resources. For specific application requirements, these services have to be tailored to meet the particular needs.

Many service discovery protocols have been proposed. Examples include UPnP, SLP, and Jini [82, 87, 181]. They allow automatic detection of hardware devices, software, and other resources in a computing environment along with services offered by various kinds of entities. Existing service discovery protocols differ in the way service discovery is performed, and no single protocol is suitable for all the environment. Due to the differences in the service discovery approaches implemented, a user working in one environment may not be able to search for the services available in another environment that suits the user's need. To support service discovery across different environments, new techniques are needed to integrate or bridge the service discovery protocols used in a diversity of environments.

After getting the service information, the next step is to access these services (interaction). In the network environment, the interaction between software entities is achieved by middleware. However, traditional middleware systems mostly originate from a static, closed, and controllable environment. Thus, considering service interaction in the open environment, these interaction approaches can hardly satisfy software coordination requirements due to their insufficient support in multi-mode interaction and dynamic function customization. Through the analysis of multiple popular middleware models, such as Remote Procedure Call (RPC) [26], Common Object Request Broker Architecture (CORBA) [14], Jini [181], etc., it is shown that middleware includes two facts: one is the programming model used to express the interaction manner such as objects binding, remote method invocation, etc., and the other one is the runtime support mechanism implementing entity interaction such as a group of proxies, including the stub, the proxy, and the skeleton.

From the perspective of the open environment, there are two shortcomings of traditional models: one is the rigidity of interaction mode. In other words, one model can usually support only one type of interaction mode. It could be either tightly-coupled RPC/RMI or loosely-coupled interaction based on shared space. The other shortcoming is the rigidity of the function of proxy, which means that the modality and function of the proxies in the middleware are usually fixed and could not be adapted according to the varieties of application system requirements. In the open network environment, we hope to achieve the following two goals. First, we can design and customize the interaction mode according to specific environment conditions and application system requirements, so as to support not only the multi-mode interaction, but also the coexistence of various interaction modes in one application system. Second, the form and function of the proxies in the middleware support system can be generalized, so that we can specify these proxies according to the customized interaction mode and tailor corresponding software entities according to the demands of the interaction modes.

The above observations motivate us to inspect the common set of popular service discovery protocols and extract them into a unified adaptor. It has two major parts, *i.e.*, adaptor primitives and specific adaptor mapping which are encapsulated in a connector. Combined with the context information, the client uses the primitives to communicate to the service provider in the heterogeneous environment adaptively. For interaction, the coordination logic is encapsulated as a series of interceptors. According to the user's interaction requirements, a combination of interceptors of the connector can be adaptively formed and thus multi-mode interaction is achieved.

6.2   Approach

6.2.1   Context model and system architecture

As explained in previous chapters, an ontology based context model can give a common understanding of the items in an open environment. This is extremely important for adaptive service discovery and interaction in the presence of heterogeneity. Considering the fast evolution of software and hardware technology, it is important that decisions regarding the context for service discovery and interaction are extensible. Thus, we remain as conservative as possible, keeping open the options for changes in the context model. We opted to define a basic, generic ontology model. Examining the discovery and interaction process, we decided to build our ontology around four main entities: *i.e.*, the platform, the service, the user, and the network.

Platform. Platform denotes the hardware and software descriptions of a specific device. This includes specifications of the processor, available memory and bandwidth, as well as, information about the operating system and other available

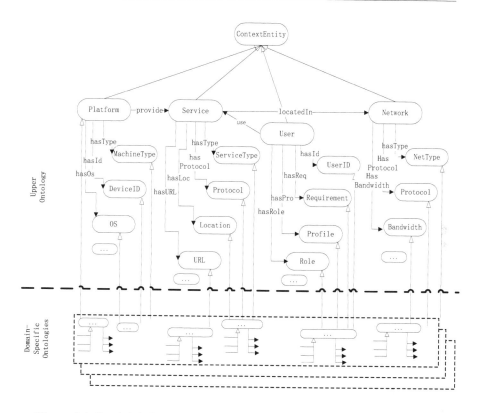

Figure 6.1: Partial definition of the context ontology for service interaction

software libraries. *Service*. Services provide specific functionality to the user. Descriptions of services include their identities, discovery protocols, addresses, interaction modes and types, etc. *User*. Users have particular requirements on the services and the interaction modes. It includes their profiles, preferences, etc. For their inter-relations and users access services, service may also acquire other services to accomplish a particular task. Platforms provide services and these conversations take place in the *network*. We divide the context model into upper ontology and domain-specific ontology. The upper ontology captures general features of basic contextual entities; while the domain-specific ontology describes the entities in detail. Figure 6.1 illustrates the partial definition of the ontology for service interaction.

The system architecture is described in Figure 6.2. It consists of three major parts, *i.e.*, connector, context information, and services. The connector is composed of some service discovery primitives, specific mappings, and a series of interceptors. The discovery manager and interaction manager are responsible for connecting the primitives and protocol-specific mappings according to the context information. We define an *environment* as a service discovery and interaction

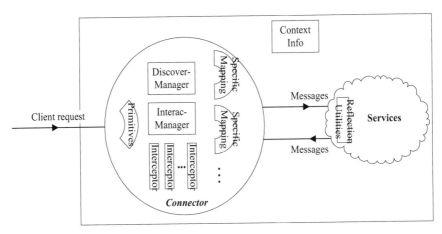

Figure 6.2: Architecture for service discovery and interaction

domain, where a native service discovery service system is able to find a specified resource if they are available in the domain. A native service discovery system can take the form of an existing service discovery protocol, *e.g.*, SLP, or Jini. Moreover, it can also be a tailor-made mechanism, *e.g.*, ReMMoC [83] which supports the inter-operability of service discovery within an environment.

6.2.2   *Service discovery primitives and mapping*

We studied existing service discovery protocols and observed that all the service discovery systems provide the similar functionality for end users and differ only in the underlying models and operations. We can abstract the common characteristics of existing approaches and provide a universal set of primitives that enable service discovery across diverse environments. The format of the proposed primitives is the following:

[Return_Values] Primitive_Name [Parameters].

Primitive_Name represents the functionality requested by the client. *Parameters* are a set of attributes required in the discovery process. *Return_Values* is a set of values returned to the client. The discovery process starts when the client expresses his interest in a particular service. Next, the discovery system searches for the service and returns results to the client. The common feature of all the existing service discovery mechanisms is that they support the process of looking for service and enable access to the service. Therefore, we propose two universal primitives: Discovery and Access. Specific adaptor primitives with parameters are shown in Figure 6.3.

The Discovery primitive is of the following format:

[RV_ST, RV_F, RV_S] Discovery (Service Type, Filter, Security)

A client requests the specific service by providing its type in the ServiceType

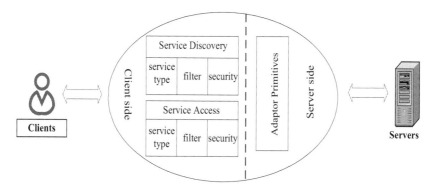

Figure 6.3: Service adaptor primitives

parameter. The attributes of the service are specified by the client in the Filter parameter. If the authentication of the client for using the service is required, the client provides the required information in the Security parameter. The Discovery primitive provides three return values. *RV_ST* is the return value containing list of the discovered services of the requested type. Each of the discovered services has a unique ID assigned by the discovery manager. *RV_F* provides a list of attributes for each of the discovered services. For example, for a printing service, an attribute can be specified to indicate whether the printer supports color printing or only black&white. *RV_S* is a return value indicating the authentication status. After the successful discovery of the requested service, the client selects one of the returned services and access the service by using the Access primitive, which is of the following format:

[RV_status] Access (ServiceID, Service Method, Attributes)

The client provides the parameter ServiceID, the ID of the selected service. The *ServiceMethod* parameter specifies the specific method provided by the service. The *Attributes* parameter is a set of attributes specific to the particular service and users' interaction properties. *RV_status* is the return value indicating the status of accessing the service, *e.g.*, success or failure. Details about the multimode interaction support will be discussed in the next section.

Adaptor mapping performs the mapping from adaptor primitives to the specific commands used in the target environment. As we mentioned before, there is a number of currently existing service discovery mechanisms, including protocols and inter-operability systems. Moreover, new mechanisms are under development. Therefore, the many-to-many approach providing a mapping between the primitives of each pair of these mechanisms would result in a heavy system with difficulties in adapting to the future service discovery mechanisms. Our approach is a one-to-many approach, providing the tailored mapping from the adaptor primitives to each native protocol used in the target environment.

For the structure of components of the adaptor mapping, we refer to Fig-

Table 6.1: Function of adaptor mapping components

Component	Function
ESD Description	Rules specifying the characteristics of the service discovery protocol used in the particular environment, in the form of the primitives extracted from the native protocol
Translator	Module that translates adaptor primitives into native primitives, and outputs an algorithm for service discovery using the native primitives
ESDA	Agent that uses the algorithm generated by Translator to perform service discovery in the target environment. Depending on the native service discovery protocol, ESDA can register for service advertisements or performs active service discovery

ure 6.4. More specifically, the functions of mapping components are described in Table 6.1. *Translator* performs mapping based on the *ESD Description* (Environment Service Discovery Description), which is a set of rules specifying the characteristics of the service discovery protocol used in the particular environment. Because different environments adopt diverse models at the low level, the rules used by *ESD Description* are tailored to each environment and are described in the form of the primitives extracted from the native service discovery protocol. *Translator* outputs an algorithm for the service discovery using the native primitives. *ESDA* (Environment Service Discovery Agent) is representing the client in the process of discovery in the target environment. It uses the algorithm generated by Translator. From the server's point of view, *ESDA* is the same as any other client in the environment using the native protocol for service discovery. All communications take place between the service discovery protocol and *ESDA*. Another function of *ESDA* is to provide the client with the return values from the environment, *e.g.*, information about the attributes.

6.2.3   *Interaction programming model*

There are many existing interaction modes. Among them we are familiar with the RPC mode, the Tuple-Space mode [47], and the Meeting mode [139]. Supported by traditional middleware, the software entities at both sides of the server and the client usually share one interaction mode; furthermore, they are only able to interact with other entities which adopt the same interaction mode as themselves. In this way, the services can only interact in a single mode and cannot have interaction with others adopting a different mode in the same software system.

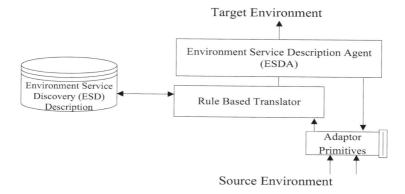

Figure 6.4: Service adaptor mapping

However, with the emergence and development of services in the open and dynamic network environment, network applications tend to be integrated with the services which are based on the idea of 'service first, interaction second'. This idea advocates the principle of 'separation of concerns' [61, 141]. It requires the service design and development to be independent from the interaction mode, emphasizes the configuration of the interaction mode in service interconnection, embodies various environment requirements through interaction programming, and at last, supports the multi-mode interaction of the service.

There are several ways to support multi-mode interaction. One method is to make an 'enumerative match' among the existing interaction modes. It means to design a switch for every two interaction modes. Obviously, this method involves a lot of effort since the number of the switches might be enormous and the handling of the newly added interaction mode is very complex. Another method is to adopt the principle of 'separation of concerns'. We can analyze the structure of the interaction modes, decompose the basic components of them, and give a uniformed interaction programming model following the principle of consistency and completeness. Through the investigation of the conversion of one interaction mode to others, we can program crossing various interaction modes during service integration period, so that the services following different interaction modes can achieve multi-mode interaction.

Service requestor, provider, and the interaction media constitute three main features of one interaction. In detail, the above features are orthogonal and can be decomposed into sub-features until atomic features. For example, the requestor element includes service specification, non-functional constraints, service tailor, and so on. In [173], we proposed a decomposition model of interaction features. Since we focus on entity configuration, we adopt a multi-dimension model as Figure 6.5 shows. The integrator could refer to this hierarchical interaction features configuration model to interconnect services. According to application requirements and environment conditions, through the configuration of the inter-

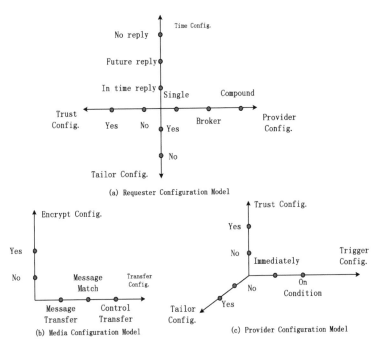

Figure 6.5: Interaction features decomposition model. Reprinted from [173], with permission from Springer Science+Business Media.

action requestor, provider, and media, the interaction mode is configurable, and the software entity interconnection is programmable.

The definition framework of interaction mode is shown in Listing 6.1. Actually, one interaction mode is constituted by one or more interaction 'facets'. Each 'facet' includes one or more compound feature's. Each compound 'feature' is composed of one or more compound/atomic features. Atomic features refer to those that cannot be decomposed and can directly express interaction features, such as name, type, parameters, and corresponding handling methods. For example, one interaction which is constrained to return result in one second can be expressed by $< TimeLimit, int, 1, TimingProc >$. TimingProc in this expression refers to a temporal application which will be provided by interaction decomposition programmers.

Based on the XML Schema definition, we propose a hierarchical and multi-dimension interaction features decomposition/configuration model. It is a reference model during service integration. Aided by visual development tools, we can decompose, organize, configure, synthesize, and describe interaction features. Actually, this model does not enumerate all possible facets and features because of two main reasons. First, decomposition is uncertain to some degree. The classification to facets and features is not unique. Second, this set is not a finite one. As the understanding of the interaction becomes more mature and application requirements are constantly changing, the facets and features tend to increase. That is to say, complete enumeration is very difficult and even impossible. Therefore, we try to give an open and extensible decomposition model which is based on interaction feature decomposition concept and multi-dimension space. In this way, new features can be easily added to the model by adding a new dimension or new value in an existing dimension. Based on the above analysis, some basic modeling principles of the decomposition/configuration model can be given as follows:

1. Openness: the structure of the model must be open. It should be able to describe various features and their composition and support the joining of new features.

2. Uniformity: the model must keep uniform. It should be capable of expressing various interaction features and defining existent and new interaction modes by configuration and combination.

3. Consistency: the classification of the features must be orthogonal and agree with the principle of 'separation of concerns'. The parameters of all features must be consistent and cannot destroy the semantic demands of the corresponding interaction mode.

4. Integrity: the integrity of the features and their parameters, as well as, the semantic requirements of the interaction mode, must be satisfied in the interaction mode programming which is done by the combination of the features and the configuration of the parameters.

In the requestor configuration model, the service type configuration points out the service type and gives concrete configuration suggestions in the requestor/provider relation; temporal configuration shows the synchronous/asynchronous feature in the interaction; trust configuration demonstrates whether the requestor needs trust computing assurance from the provider (if needed, the integrator has to specify the trust computing evaluation policy); service customization configuration indicates whether it is necessary to customize the services of the provider in the interaction (if required, the integrator should specify the customization configuration).

In the media configuration model, information model configuration specifies whether the media transfers messages as message passing, or matches messages as space (if it does, the integrator must specify the matching rules) or transfers control as agent (if it is agent, an execution agent will be created by the environment at runtime); the channel encryption configuration indicates whether the interaction messages are encrypted.

In the provider configuration model, the trigger method configuration shows whether the provider is able to provide services after receiving the request (if it does, the integrator has to specify the trigger policy); trust request configuration indicates whether the provider needs trust computing from the requestor (if it does, the integrator must give the trust assurance policy); service customization configuration specifies whether the service functions need to be customized (if it does, the integrator should indicate the customization policy).

Listing 6.1: Interaction mode definition

```
<?xml version="1.0" encoding="UTF-8"?>
<xs:schema xmlns:xs=http://www.w3.org/2001/XMLSchema
 xmlns="http://ics.nju.edu.cn/interactionMode/schema"
 targetNamespace="http://ics.nju.edu.cn/interactionMode/schema"
 elementFormDefault="qualified" attributeFormDefault="unqualified">
 <xs:element name="InteractionMode"> <xs:complexType><xs:sequence>
  <xs:element ref="Aspect" minOccurs="1" maxOccurs="unbounded"/>
 </xs:sequence> </xs:complexType></xs:element>
 <xs:element name="Aspect"> <xs:complexType><xs:sequence>
  <xs:element ref="CompoundFeature" minOccurs="1"
  maxOccurs="unbounded"/>
 </xs:sequence> </xs:complexType></xs:element>
  <xs:element name="CompoundFeature"> <xs:complexType><xs:sequence>
  <xs:elementref="Feature" minOccurs="1" maxOccurs="unbounded"/>
  <xs:elementref="CompoundFeature" minOccurs="1"
  maxOccurs="unbounded"/>
 </xs:sequence> </xs:complexType></xs:element>
 <xs:elementname="Feature"> <xs:complexType>
 <xs:attribute name="Name" type="xs:string"/>
 <xs:attribute name="Type" type="xs:string"/>
 <xs:attribute name="Parameter" type="xs:string"/>
 <xs:attribute name="FeatureProcedure" type="xs:string"/>
</xs:complexType></xs:element> </xs:schema>
```

From the perspective of service interaction programming, service integrators can: 1) decompose interaction mode under the guidance of above decomposition/configuration model, 2) program each feature by configuration, and 3) define feature contents and corresponding handling method. All of the above configuration and policy information will be collected by the integration environment and handled by the runtime environment at runtime. In this way, we can achieve adaptive multi-mode interaction.

6.2.4 Interceptor based multi-mode interaction

From the process of interaction feature decomposition and synthesis, interaction mode is determined by feature configuration items selection, parameter, and handling process specification. Since configuration items, parameters, and processes are independent of service providers, the coordination media of service cannot obtain concrete parameters and handling methods before service execution. That is to say, all these features should be configured before service running. Thus, according to each service, the service interaction mode programming information and related handling program will be encapsulated into connectors. These con-

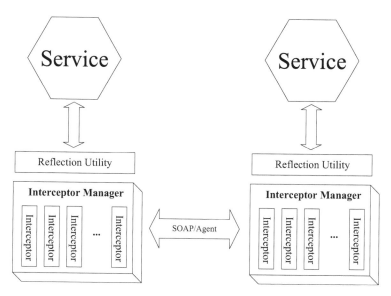

Figure 6.6: Conceptual framework for multi-mode interaction

nectors together with related components are further reified into an architectural agent. For details of the runtime architectural agent, we refer interested readers to our previous work [121]. The agent will be dispatched to each node where the service being integrated lies. One service is mapped to one agent. Actually, services do not directly interact with other services. All interaction and coordination behaviors are controlled by the agent which encapsulates interaction mode information. The service only receives the commands from its agent and carries out its own function. From this concept, all brokers which play an important role in each interaction mode (*e.g.*, Stub/Skeleton in CORBA, Space in Tuple Space, and Message Queue in Message middleware) can be regarded as specific cases of an agent. Thus, multiple agents construct a fine interaction environment to hide the heterogeneity of heterogeneous services, and finally provide a uniform framework for multi-mode interaction among services. The framework is illustrated in Figure 6.6.

The implementation of programmable coordination media which supports multi-mode interaction may adopt the reflection technique. Meta-interface of coordination media is added, through which we could change the behavior of coordination media dynamically. The detailed design idea is presented as follows:

1. The interaction element configuration and its parameters are reified into unitary data objects while the processing rule of the parameters is reified into meta-programs. The meta-program is uploaded to the coordination media through meta-interface and becomes the interpreter for meta-data objects in coordination media.

2. The coordination media intercepts the above data objects representing concrete configuration of the interaction through meta-interface and operates these objects through the execution of the interpreter. During this process, a new group of interaction data objects might be generated and submitted to the next interpreter.

3. These interpreters of configuration item parameters are organized based on the Interceptor design pattern [73]. Sequential execution of the interceptors realizes the combination of multiple configuration items of the interaction. The Interceptor is uploaded dynamically. The structure of interceptor mode limits its access to the resources in the server, and addresses the security problems arisen from uploading.

Based on the above analysis, we designed and implemented a programmable coordination media PCM (Programmable Coordination Media) [173]. As described previously, the agent encapsulates all interaction features and handling methods. Each feature handling method will be reified as an Interceptor. When the agent migrates to the node where the service resides, PCM writes the agent's interceptor configurations into Interceptor Manager during the deployment phase. The configuration is processed sequentially based on the interaction identity during runtime.

The PCM architecture is described in Figure 6.7. As the core part of a coordination application, PCM exists in the application server in the form of singleton instance. Its life-cycle fully depends on the life-cycle of the host application server. Interceptors in PCM rely on the dynamic arrivals of the agents. During runtime, service proxy provides simple and consistent message interfaces for client applications and services. The requests from client software are encapsulated by service proxy and then intercepted by the PCM proxy. PCM proxy passes the requests to *Interceptor Manager* after identifying them. *Interceptor Manager* would process the request objects sequentially according to the configuration, generate a soap message or an execution agent which is used to transfer the control, and then send them to PCM at the server side.

6.2.5   Adaptation

Service discovery and interaction managers constitute the most important part of the adaptation. They connect with the context model, the protocol specific mapping modules, and interceptors. They have a common module responsible for querying the context ontology. The service discovery manager mainly searches for the available services and their protocols in their environment. The service interaction manager is mainly interested in the user's interaction requirements.

For example, when the client has a request for the weather forecast service and issues a message to the service discovery manager. The service discovery manager contacts the local registry center. The registry center returns a service

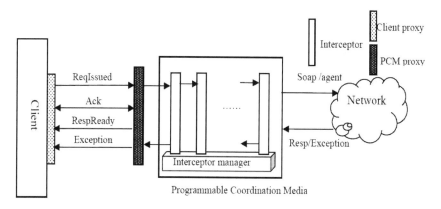

Figure 6.7: PCM architecture. Reprinted from [173], with permission from Springer Science+Business Media.

provider and feeds the result back to the client. The result includes the service URL and the protocol. Normally, the client interacts with the service through the interceptors. This is the ideal case. However, as the environment is dynamic, it is possible that the service responds very slowly. The user must specify some rules to adapt to these conditions. However, if the similar service is found in a heterogeneous environment, the client may request the interaction should be securely processed. This requires a flexible combination of the adaptors and interceptors. As these concerns are separated from the business logic and encapsulated in the connector, the runtime adaptation can be achieved. Here, the excerpt of the query language and automatically generated interceptor configuration file are illustrated in Listing 6.2 and Listing 6.3, respectively.

Listing 6.2: Service information query

```
PREFIX ndadapt:    <http://ics.nju.edu.cn/ndadapt>
SELECT ?serviceName ?serviceURL ?serviceProtocol
WHERE { ?x ndadapt:hasName ?serviceName.
        ?x ndadapt:hasURL ?serviceURL.
        ?x ndadapt:hasProtocol ?serviceProtocol.
}
```

We use SPARQL [145] as the query language for its generality and tool support. Generally, the query will return a set of available service providers. They will be arranged into a list. The nearest will be considered as a first priority, and the rest will be considered as backups.

To adapt to the context changes, the system should have the reasoning ability based upon the user-defined rules. We use Jena [123] as the underlying reasoning engine. For this example, we can specify the rules as in Listing 6.4.

Listing 6.4 describes the replace rules defined by the Jena rule format. It means that when the response time of the service is slower than the threshold (3000 milliseconds in the example), and there is another compatible service which is free, then another service will replace the current one.

Listing 6.3: Interceptor configuration

```xml
<?xml version="1.0" encoding="UTF-8"?>
<connectors>
    <connector name="Connector0">
      <interceptor-tree name="Caller">
        <interceptor
          className="SynMethodInterceptor"/>
        <interceptor
          className="DigestClientInterceptor"/>
        <interceptor
          className="InvokerInterceptor"/>
      </interceptor-tree>
      <interceptor-tree name="Callee">
        <interceptor
          className="DigestServerInterceptor"/>
      </interceptor-tree>
    </connector>
</connectors>
```

Listing 6.4: Service replace rule

```
@prefix ndadapt:<http://ics.nju.edu.cn/ndadapt>
@prefix xsd:<http://www.w3.org/2001/XMLSchema#>
[r1:(?id, rdf:type, ndadapt:service), (?responseTime, ndadapt:intValue, ?value),
  greaterThan(?value, '3000'^^xsd:int)->(?id, ndadapt:state, "slow")];
[r2:(?id1, rdf:type, ndadapt:service), (?id2, rdf:type, ndadapt:service),
  (?id1, ndadapt:descriptor, ?descriptor1), (?id2, ndadapt:descriptor, ?descriptor2),
  equal(?descriptor1, ?descriptor2)->(?id1, ndadapt:compatible, ?id2)];
[r3:(?id1, rdf:type, ndadapt:service), (?id2, rdf:type, ndadapt:service),
  (?id1, ndadapt:compatible, ?id2), (?id1, adapt:state, "slow"),
  (?id2, ndadapt:state, "free")->(?id2, ndadapt:replace, ?id1)]
```

Listing 6.3 illustrates the interceptor configuration file for the interaction between the client and the service provider. The connector combines three interceptors, *i.e.*, *SynMethodInterceptor*, *DigestClientInterceptor*, and *InvokerInterceptor*. They are responsible for the synchronization, digital signature, and normal invocation, respectively. The interceptor configuration file will be generated automatically by the interceptor manager. The wrapper module will serialize the corresponding interceptor classes and send them to the PCM on the server side.

6.3 Performance Analysis

We have developed a prototype of the proposed approach. The system currently has two service discovery environments (*i.e.*, Jini, SLP) and supports multi-mode interaction. For Jini, we use Jini2_1 [181] which is an implementation of the Jini technology from Sun Microsystems. For SLP, we use a pure Java implementation of SLP-jSLP. A PCM module is implemented which stays in the client and server sides. They communicate through SOAP. We also set up a simple context database for service query.

As an example, we implemented a weather information service using Jini

Table 6.2: Discovery overhead experiment

	In Jini Environment	In SLP Environment
Service discovery time without adaptor [ms]	638	152
Service discovery time with adaptor [ms]	743	197
Overhead [%]	16	30

and SLP protocols separately. Based on the scenarios discussed in the last section, the process includes the service discovery, interaction, and adaptation. The soap message contains the particular interceptor information. To evaluate the performance of the proposed approach, first we measured the time cost of service discovery in the two environments using native service discovery clients (both for Jini and SLP). Next, we compared that with the discovery time using the adaptors, respectively. Experiments are conducted with a desktop computer with CPU Pentium D 2.8GHz, 1G memory and a laptop computer with CPU Pentium Mobile 1.6GHz, 512M Memory connecting to the LAN with 100Mbps Ethernet and 11Mbps 802.11b protocol, respectively. The results of the experiments are presented in Table 6.2. In the Jini environment, the average service discovery time is 638 milliseconds, and in the case of using an adaptor, the average time is 743 milliseconds. In the SLP environment, the average service discovery time is 152 milliseconds. The overhead that the adaptor imposes on service discovery time is mainly introduced by the socket setup and request parse. In the experiment, we did not consider the time cost introduced by the context query. To measure the interaction cost, we conducted three types of invocations of the services. One type is without the interceptors. The second type is through empty interceptors. The third type is through a customized interceptor. In the example, we added an encryption interceptor. We calculated the overhead introduced by the reflection. For each type, we tested 30 times and calculated the average value. The result is illustrated in Table 6.3. The average overhead is around 15% introduced by the reflection. The added interceptor overhead is around 26%. For adaptation process, we calculated the reasoning time and the automated reconfiguration time of the connectors. The average reasoning time in our experiment is around 1130 milliseconds, while the reconfiguration of the connector takes around 1342 milliseconds.

6.4 Related Work

Many service discovery protocols have been proposed for different applications in various environments. Survey and comparisons between the existing service

Table 6.3: Interaction overhead experiment

	In Jini Environment	In SLP Environment
Native invocation [ms]	865	342
Interaction with native interceptor [ms]	990	402
Interaction with one interceptor [ms]	1223	518

discovery protocols can be found in [23, 150, 199]. New service discovery protocols targeted at pervasive environment are also emerging [103]. In this section, we focus on the works that support inter-operability of the service discovery protocols. Several approaches on supporting multi-protocol service discovery have been proposed [56, 83, 108].

All the existing service discovery mechanisms realize the concept of client/server application. Clients are entities that need some functionality (service), and servers are entities existing in the environment and offering this functionality. Service discovery is the framework for connecting clients with services. Existing work towards achieving multi-protocol service discovery uses a middleware approach but differs in where the inter-operability support is provided. The middleware can be on the service side, the client side, or the intermediate entity.

An example of providing the support on the service side is INDISS [29] designed for home networks. INDISS provides parsers and composers, which decompose a request from the source service discovery protocol into a set of events and then compose them into a message understood by the target service discovery protocol.

INDISS can also be deployed on the client side. Another example of middleware on the client side is ReMMoC [83]. In ReMMoC, the client side framework provides the mappings between all the supported service discovery mechanisms. Abstraction of discovery protocols to the generic service discovery is achieved by using a generic API or doing discovery transparently to the client. In this approach, all possible mappings need to be provided at the client side. Service side support can also be based on service proxies [103]. When a new service appears or disappears in one of the environments, the framework detects it and creates or removes its proxies from another environment. However, this approach requires dynamic service proxies to be implemented for each environment, which would increase the developer's workload.

Another way to provide inter-operability support is to provide an intermediate entity [108, 150]. In Open Service Gateway Initiative (OSGi) [56] all the connections and communications between the devices are brokered by a Java-based platform. An internal service registry exists in the framework. Interoperability is achieved by providing an API to map the given service discovery protocol to OSGi and vice versa. Supporting new service discovery protocols requires defining new application programming interfaces (APIs) for them. Gateway functionality is also utilized by protocol adapters in the FuegoCore Service broker [108] designed for mobile computing environment. The service broker registers the mappings between its internal template and the external templates used by different service discovery protocols. Extending the FuegoCore service broker to new service discovery protocols requires creating and deploying additional service discovery protocol adapters. INDISS [29], mentioned above, can be deployed as an intermediate entity as well. The intermediate entity approach requires the broker to integrate all the adapters into one system. In a network with a large number of service discovery protocols the framework may not be scalable.

Another approach is to provide service that discovers and interacts services across different environments [149]. In MUSDAC, it registers itself in all environments, so clients can use whatever protocol to discover it. However, clients must have the knowledge about MUSDAC and the process of discovering the service has high processing requirements. The centralized approach has the scalability problem. In the future pervasive environment, an unlimited number of service discovery mechanisms will emerge. Providing all possible translations in one middleware would result in a very heavy system. The work reported in this chapter is based on the analysis of the following requirements on the interoperability system for pervasive computing: 1) no change should be imposed on the existing service discovery mechanisms; 2) no functionality of the environment should be compromised; 3) the system should be lightweight, scalable, and extendible; 4) multi-mode service interaction should be supported; 5) both standard and tailor-made service discovery mechanisms should be supported. Our approach addresses all of these requirements. We also incorporate the semantic context modeling techniques to describe the environment and the related reasoning rules. Therefore a self-adaptive service discovery and multi-mode interaction mechanism can be realized.

6.5 Summary

This chapter presents a connector-based approach for service discovery and multi-mode interaction in a heterogeneous environment. The mechanism contains service discovery primitives, protocol-specific adaptors, and a series of interceptors. Based on the context ontology and rule reasoning, the flexible combinations of these elements can support heterogeneous service discovery and multi-mode interaction. Our contribution to the inter-operability of service discovery is providing a solution in a lightweight manner. It bridges not only all existing but also future service discovery systems, including the standard ones, as well as, service discovery mechanisms that support multiple protocols within the domain.

The separation of the coordination from the computation enables the flexible interaction modes.

We focus on the adaptation mechanism, based on flexible combinations of interceptors and connectors. However, we did not develop a full-fledged system for automated semantic-based service discovery and matching. Moreover, operating in a heterogeneous environment involves the security risks which are not addressed in our current approach.

FORMAL MODELING AND ANALYSIS

Chapter 7

Adaptation Rules Conflict Detection

CONTENTS

7.1 Introduction

Architecture-based runtime adaptation has been proposed as an effective way to ensure the continuous availability of component-based software systems in an uncertain environment [136, 137]. Generally, such systems have embedded sensors to probe the context information and can propagate the changes to the rule management modules and direct the corresponding dynamic architectural reconfigurations accordingly.

In decentralized and distributed settings, interesting context changes might be reported from place to place, thus, it is highly possible that multiple architectural reconfiguration rules are applicable at the same time. These rules usually address different aspects of concerns, either functional or non-functional, and are not necessarily independent of each other. So, applying one rule may disable another one. Therefore, in this case, how to detect the potential conflicts between the reconfigurations becomes an important issue for efficient adaptation management.

To formalize the problem, given a set of reconfigurations: $S_r = \{r_1, r_2, \ldots, r_n\}$, we need to find the tuple sets:

$$\{(r_i, r_j) \mid r_i, r_j \in S_r \wedge (DEP(r_i, r_j) = TRUE \vee CONFLICTS(r_i, r_j) = TRUE)\}.$$

For example, in a component-based software system, the trust management module reports a fraud behavior of a third-party component and the adaptation rule indicates its removal from the system. Meanwhile, a performance monitor reports a low-performance event of another component and the adaptation rule implies an addition of a new back-up component. If the new component relies on the functions of the third-party component stated above, a conflict would happen between these two reconfigurations. Manually assigning each rule with an application-specific priority as commonly adopted leads to an ad-hoc solution which only partly addresses the problem, as the developer may still be unaware of the potential conflicts or the dependency relations between the involved reconfiguration possibilities. Besides, this mechanism lacks formal treatment of such problems.

Graph notations have been widely applied to software architecture modeling [18, 28, 115] due to their unique characteristics. Firstly, graph representation has the advantage of an intuitive visual correspondence to the common practice of box-line notation of software architecture; secondly, the widely recognized importance of the connector as a first-class entity can be well captured by the notion of edge attributes; thirdly, by *production*, the graph rewriting techniques can be employed to describe the evolution and to check some interesting properties, such as confluence, conflicts, etc. formally.

In this chapter, we will focus on the third point listed above and inspect the applicability of in the domain of dynamic software architectural reconfigurations. We leverage the theories from the critical pair analysis, inspect the relations between the reconfiguration rules such as dependence and confluence, and present a formal approach to analyze the potential conflicts of multiple reconfigurations. The rest of the chapter is organized as follows. Section 7.2 introduces some basic notions relevant to the formalism of conflict analysis and then outlines our approach. To illustrate its application and feasibility, a case study is presented with tool support in Section 7.3, followed by some discussions on the lessons learned and the limitations of our approach in Section 7.4. Section 7.5 reviews some related work. Section 7.6 summarizes this chapter.

7.2   Reconfiguration Modeling and Analysis

For architecture-based self-adaptation rules, basically, they have a spectrum of two layers as in [110], *i.e., goal management layer* and *change management layer*. The first handles the context changes which are usually reflected in the condition part of the rule. It is on the application level. The second deals with the reconfiguration actions which are on the architecture level. For example, a load-balancing rule specifies that when the workload of a component reaches a certain threshold, a backup component will be added dynamically to the system. As the condition part depends on the concrete requirements, it varies across applications. In contrast to this, the architectural reconfigurations are manipulated at the level of constituent components and connectors, which can be specified as the graph transformation rule [184]. With such considerations, we will focus on the architecture level rule modeling. It fits in the *change actions management layer* in [110]. To detect reconfiguration dependency and conflicts of dynamic software architecture, their independent relations need to be modeled and analyzed. In this section, we will first introduce some basic concepts, and then present our approach.

7.2.1   Critical pair and dependency

Critical pair analysis originated from term rewriting and later was generalized to graph transformation to check whether the transformation system is confluent [143]. Given a graph G, a production $p : L \Rightarrow R$ and a match $m : L \Rightarrow G$, the transformation from G to H is specified as $t : G \overset{p,m}{\Rightarrow} H$. A critical pair is a pair of transformations (p_1, p_2) in which $t_1 : G \overset{p_1,m_1}{\Rightarrow} H_1$ and $t_2 : G \overset{p_2,m_2}{\Rightarrow} H_2$ have potential conflicts with the minimal $G = m_1(L_1) \cup m_2(L_2)$ to ensure that the critical pair sets for two rules are finite. Intuitively, for two rules both applicable to a graph G, to become a critical pair, at least one rule deletes the common item(s) of their match images in G or generates graph objects that violate the application conditions, *e.g.*, NAC [88], of the other one.

To analyze critical pairs and dependency, first we need to distinguish the categories of transformation independence. There are two cases for independence, *i.e., parallel independence* and *sequential independence* [154]. Given two *alternative* direct transformations $t_1 : G \overset{p_1,m_1}{\Rightarrow} H_1$ and $t_2 : G \overset{p_2,m_2}{\Rightarrow} H_2$, if $m_2(L_2) \cap m_1(L_1 - dom(p_1)) = \emptyset$, we say the direct transformation p_2 is of p_1. If both t_1 and t_2 are mutually weakly independent of each other, the two transformations are parallel independent. In case of two *consecutive* direct transformations $t_1 : G \overset{p_1,m_1}{\Rightarrow} H_1$ and $t_2' : H_1 \overset{p_2,m_2'}{\Rightarrow} X$, we say that t_2' is sequentially independent of t_1 if $m_2'(L_2) \cap m_1(R_1 - p_1(L_1)) = \emptyset$ and $m_2'(L_2 - dom(p_2)) \cap m_1(R_1) = \emptyset$. According to the local Church-Rosser theorem [154], the parallel independence of p_1 and p_2 is equivalent to the statement that there is a graph X and direct transformations $H_1 \overset{p_2,m_2'}{\Rightarrow} X$ and $H_2 \overset{p_1,m_1'}{\Rightarrow} X$ such that $G \overset{p_1,m_1}{\Rightarrow} H_1 \overset{p_2,m_2'}{\Rightarrow} X$ and $G \overset{p_2,m_2}{\Rightarrow} H_2 \overset{p_1,m_1'}{\Rightarrow} X$ are sequentially independent transformations. In the context of multiple architectural reconfiguration choices, this theorem states that, if the two (or multiple) reconfigurations are parallel independent, they can be applied in an arbitrary order, and the transformation sequences will result in the same result.

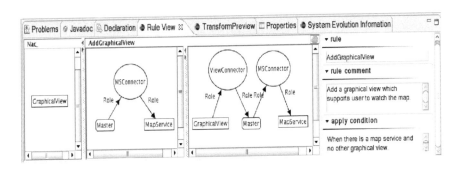

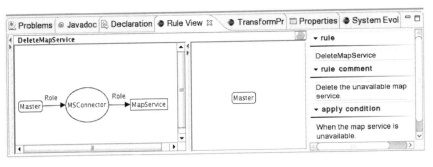

Figure 7.1: Reconfiguration conflict illustration

Independence is only a subset of the relations between transformation rules. There are other cases, such as the dependence and the conflict. Informally speaking, if the transformation result of one rule enables another one, it implies a dependency. If the application of one rule disables another one, the two rules represent a potential conflict and form a critical pair. As a concrete example, Figure 7.1 illustrates a conflicting situation. The rule *DeleteMapService* denotes a detachment of the map service from the master component. This may happen when the map service component is unavailable or untrustworthy. But the second rule *AddGraphicalView* adds a graphical interface component to display the map data retrieved from the map data component by the master component. This rule may be invoked when the system developer wants to provide the value-added service or when the network bandwidth is in a good state. If the conditions of both reconfiguration rules are satisfied, applying the *DeleteMapService* rule first would disable the application of the other one, as the left side of the rule *Add-GraphicalVew* requires the existence of the *MapService* component. However, if the rule *AddGraphicalView* is applied first, it will not disable the application of *DeleteMapService*. Therefore, there is an asymmetric conflict in this critical pair.

7.2.2 Architectural reconfiguration analysis

We use AGG [171] to calculate the relations between graph transformation rules. AGG supports the critical pair and dependence analysis. It computes by overlapping the left hand sides of two rules in all possible ways to see whether they constitute a critical pair. AGG distinguishes three kinds of conflicts: *delete-use, produce-forbid, change-use attribute*, and three kinds of dependencies: *produce-use, delete-forbid, change-use attribute*. For details, we refer interested readers to [113]. After getting the relations between the pairs of the reconfiguration rules, we classify them into four sets, *i.e.*, *independence, symmetric conflicts, dependence*, and *asymmetric conflicts*. As dependence and asymmetric conflicts are partial order relations, if the concurrent rules belong to these two sets, the first rule of the pairs should be applied first, so as to enable the other one. If the concurrent rules are in the independence set, which is free of conflicts and dependence, the order of applying them can be set freely. If the concurrent rules are in the symmetric conflicts, applying either one will disable the other. In this case, human expertise has to be involved to decide which one should have priority.

The next step is to map the transformations to the corresponding architectural reconfiguration. As stated previously, architectural reconfigurations can be reduced to the manipulation of the constituent elements of the software architecture. We extract some primitive commands from Acme [76]. The reason why we use Acme as the underlying elementary commands lies in three considerations: 1), Acme can be used as the common denominator of existing ADLs, as it provides the most common sets of ADL design concerns; 2), Acme is a general-purpose architecture description language itself and can be extended flexibly; 3), there are program libraries which can be integrated in our tool. The basic reconfiguration commands of

Acme are: *AttachmentCreateCommand, AttachmentDeleteCommand, ComponentCreateCommand, ComponentDeleteCommand, ConnectorCreateCommand, ConnectorDeleteCommand, PortCreateCommand, PortDeleteCommand, RoleCreateCommand,* and *RoleDeleteCommand.* Accordingly, we have three graph models, *i.e.*, the component model, the connector model, and the role model. The port and the attachment are implemented as attributes of the component model and the role model, respectively.

Algorithm 1 explains how to generate the primitive reconfiguration command list from the transformation rules. It iterates the sets of graph elements which have been deleted and created by the transformation. The sets are parsed by AGG together with their attributes, such as their labels, locations, interface names, properties, connector types, and other application specific descriptors. As this step is conducted after the relation analysis of the transformation rules, the generated command list is free of conflicts. Therefore, it can be applied to the underlying architecture model. The reason why we iterate twice for each set is to guarantee there is no dangling edge during the deletion or creation of the vertex.

To facilitate graph transformation-based dynamic architecture modeling and reconfiguration commands generating, we have developed an integrated tool: VIDE. AGG has been embedded as an external utility to analyze the rule relations. VIDE is implemented in Java and released as an Eclipse plug-in. It supports the graph grammar directed development of software architecture and transformation. In this way, only rule compliant transformations can be applied. It bridges the gap between our previous work on runtime architecture object [121, 194, 197] and the analysis of concurrent transformation rules. The generated command list of the rules can be applied to the architecture object model, and, in this way, the potential conflicts are eliminated before applying the rules. Figure 7.1 presents the rule view of the example transformations. A complete description of VIDE is out of the scope of this chapter, Figure 7.2 gives a basic overview of the user interface.

7.3 Case Study

In this section, we use a simple online ticket-booking system as a running example. The system normally offers online ticket booking services plus other value-added services, for example, location service, weather information services, etc. Those services are distributed and integrated from other third parties. It also has a graphical display component depending on the workload condition of the system or other factors. The type graph of the architecture elements is shown in Figure 7.3. There are three child-types of component type, *i.e.*, *master, slave,* and *view*. *Master* component is responsible for receiving the requests from the clients and delegating the work to other components. *Slave* component handles the requests from the master component and further has three different child-

Data: createdElements, deletedElements
Result: cmdList

```
1   begin
2   |    cmdList ⟵ ∅;
3   |    for m ∈ deletedElements do
4   |    |    if m instanceof RoleModel then
5   |    |    |    adc ⟵ AttachmentDeleteCommand;
6   |    |    |    rdc ⟵ RoleDeleteCommand;
7   |    |    |    adc.setAttributes; rdc.setAttributes;
8   |    |    └    cmdList ⟵ cmdList ∪ {adc} ∪ {rdc};

9   |    for m ∈ deletedElements do
10  |    |    if m instanceof ComponentModel then
11  |    |    |    cmdc ⟵ ComponentDeleteCommand;
12  |    |    |    pdc ⟵ PortDeleteCommand;
13  |    |    |    pdc.setAttributes; cmdc.setAttributes;
14  |    |    └    cmdList ⟵ cmdList ∪ {cmdc} ∪ {pdc};

15  |    |    else if m instanceof ConnectorModel then
16  |    |    |    cndc ⟵ ConnectorDeleteCommand;
17  |    |    |    cndc.setAttributes;
18  |    |    └    cmdList ⟵ cmdList ∪ {cndc};

19  |    for m ∈ createdElements do
20  |    |    if m instanceof ComponentModel then
21  |    |    |    cmcc ⟵ ComponentCreateCommand;
22  |    |    |    pcc ⟵ PortCreateCommand;
23  |    |    |    cmcc.setAttributes; pcc.setAttributes;
24  |    |    └    cmdList ⟵ cmdList ∪ {cmcc} ∪ {pcc};

25  |    |    else if m instanceof ConnectorModel then
26  |    |    |    cncc ⟵ ConnectorCreateCommand;
27  |    |    |    cncc.setAttributes;
28  |    |    └    cmdList ⟵ cmdList ∪ {cncc};

29  |    for m ∈ createdElements do
30  |    |    if m instanceof RoleModel then
31  |    |    |    rcc ⟵ RoleCreateCommand;
32  |    |    |    acc ⟵ AttachmentCreateCommand;
33  |    |    |    rcc.setAttributes; acc.setAttributes;
34  |    |    └    cmdList ⟵ cmdList ∪ {rcc} ∪ {acc};
```

Algorithm 1: Reconfiguration Commands Generation

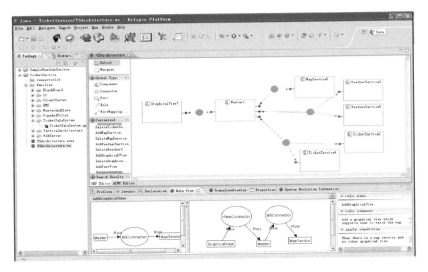

Figure 7.2: VIDE illustration

types, *i.e.*, TicketService, Mapservice, and WeatherService. *View* component is for displaying the result, either graphically or textually. The first two kinds are functional components while the last one addresses non-functional concerns. The connectors interact with the components. Corresponding to their difference, there are four kinds of connectors, *i.e.*, Viewconnector (for view service), WSConnector (for weather service), TSConnector (for ticket service), and MSConnector (for map service).

Because of the uncertainty and the changes from the dynamic environment, the system is under pressure to evolve. The developer may identify some reconfiguration patterns. For example, when a new ticket broker is active and broadcasts the online ticket-booking service, it may lead to an AddTicketService rule, while an unavailable service leads to its detachment from the system. Ticket booking and weather broadcasting are two basic services provided to customers, while the map service is a value-added service. Below, we list some common but incomplete reconfigurations.

1. *AddTicketService*: integrates a new ticket service provider. Addition may happen when a new provider is found or the requirements change to add another kind of ticket-booking service;

2. *DeleteTicketService*: detaches a ticket service component from the system. Removal may happen when the service is unavailable or the component is unwanted. There is one *positive application condition* for this rule, *i.e.*, at least one ticket service should exist after the removal;

3. *AddMapService*: integrates a map service into the system. This component

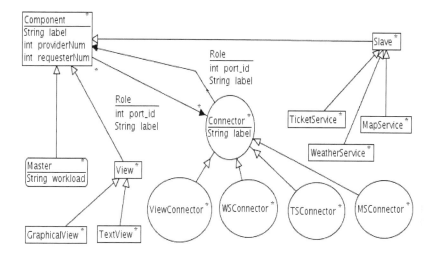

Figure 7.3: Type graph of example system

provides location services, such as spotting, traffic routing, etc. Addition may happen when the map service is available or trustworthy. In this scenario, we make an assumption that for one system there is only one map service, as the map covers whole interested cities;

4. *DeleteMapService*: detaches the map service from the system. It may happen when the component is unavailable or untrustworthy;

5. *AddWeatherService*: integrates a weather service into the system to provide weather broadcasting services;

6. *DeleteWeatherService*: detaches the weather service component from the system. Similar to the application condition of the *DeleteTicketService*, at least one weather service should exist after the removal;

7. *AddGraphicalView*: adds a graphical display to the application. This graphical interface displays the map data at the users' retrieval. Thus, graphical view component depends on the existence of the map service component. The addition may happen when the bandwidth is in good condition or the workload of the server is not heavy;

8. *DeleteGraphicalView*: detaches the graphical display component from the system. This may happen when the bandwidth is not good enough for graphical display;

9. *AddTextView*: adds a textview to the application. The textual display can save the bandwidth of the network and the workload of the `master` component;

10. *DeleteTextView*: detaches the textual display from the system. As there is only one view component at a time, this may happen when the system switches from a text view to a graphical view.

First, we examine these rules one by one and then check against the results obtained from automated tools.

AddTicketService should have no conflicting relations with other reconfigurations, since adding a ticket service component will not disable other reconfigurations. Although deleting a ticket service depends on its existence first, generally these two actions do not happen at the same time. This case is similar to other Add-Delete pairs.

DeleteTicketService should have no conflicts with other reconfigurations except itself. Thus, if multiple deletion reconfigurations of the same ticket service happen, they can be reduced to only one. This case holds similar for other Delete-* reconfigurations.

AddMapService should have a causal relation with *AddGraphicalView* and conflicts with itself, as our system assumes that only one map service is needed.

DeleteMapService should have conflicts with *AddGraphicalView*, since the graphical display is for the location spotting near the destination. The location data come from the map service component. Deleting the map service component will disable the addition of a graphical view component.

AddWeatherService and *DeleteWeatherService* are similar to the cases of *AddTicketService* and *DeleteTicketService*.

AddGraphicalView should have dependence relation with *AddMapService* and conflict with *DeleteMapService* as mentioned above. Besides, it also has conflicts with *AddTextView*, as the system assumes that there is only one view style at a time.

DeleteGraphicalView should have dependence relation with *AddTextView* since the graphical view component and the text view component are mutually exclusive. Thus, the latter reconfiguration depends on the former one.

AddTextView and *DeleteTextView* are different from the cases of *AddGraphicalView* and *DeleteGraphicalView*. This is because they are not dependent on the map service component. These two reconfigurations should be self-conflicting.

Figure 7.4 and Figure 7.5 illustrate the conflict and dependence relations analyzed by AGG. The two tables give a clear view of the relations between the reconfigurations. The obtained results correspond to our expectation. For example, *Add/DeleteTicketService* and *Add/DeleteWeatherService* have similar relationships with other reconfigurations. *AddGraphicalView* have conflicts with *AddTextView* and is dependent on *AddMapService*.

It is noteworthy that reported conflicts and dependencies are often more than actually existing ones. This is because AGG computes pairs by overlapping the left hand sides of the rules in all possible ways. For example, AGG reported 13 conflicts within two *AddGraphicalView* reconfigurations. In fact, all the reported

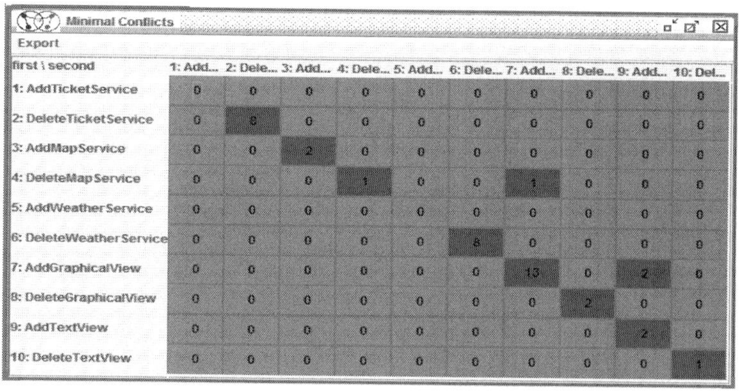

Figure 7.4: Conflict illustration

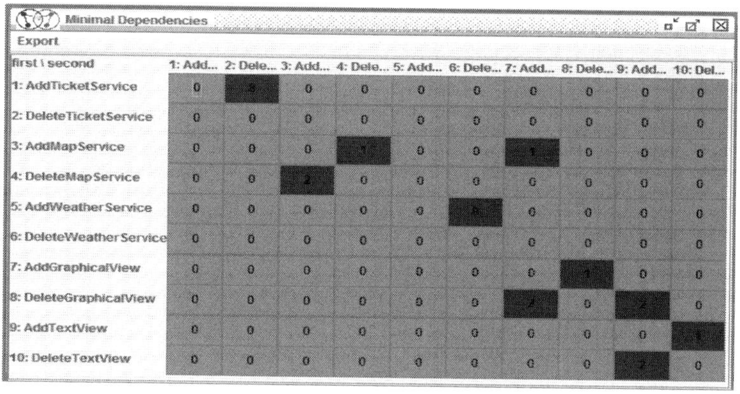

Figure 7.5: Dependency illustration

conflicts are kinds of *produce-forbid* conflict. In actual settings, it is possible that there is only one occurrence of the left hand side of the reconfiguration rule in architectural topology. This scenario is a very simplified version but still has ten reconfiguration patterns. Nevertheless, the proposed approach can capture the essential relations between the reconfigurations.

7.4 Discussion

Critical pairs only imply potential conflicts between reconfiguration rules. To decide whether or not they become real ones, other factors are needed to be taken into consideration. The key point is whether the pre-conditions will be fired at the same time.

As those reconfigurations are statically checked, the result can give the developers an insight into how to design the application specific rules. For example, developers can group all the possible reconfigurations into distinguished sets according to their inter-relations, *e.g.*, independent, dependent, and conflicting. When the number of rules increases, the proposed approach can provide a more precise and helpful guidance. Our contributions are two-fold. First we demonstrate the feasibility of the critical pair analysis based approach to modeling and detecting the potential conflicts of concurrent reconfigurations; second, we explore an application domain of conflict and dependence analysis in graph transformations.

Though we believe that our approach has potential use for the effective reconfiguration management, the approach still requires further investigations. Currently, the transformation rule is context free, and the critical pair is calculated by overlapping all the possibilities [171]. Thus, the reported number is often more than that in reality.

7.5   Related Work

In this section, we briefly overview some related work. This work is relevant in two ways: within the similar problem space, *i.e.*, the similar problem but with different solutions; within the similar solution space, *i.e.*, similar solution but in different application domains.

In [75], Garlan *et al.* presented a runtime software architecture based framework Rainbow. Rainbow uses an abstract architectural model to monitor a system's runtime properties and to evaluate the model in terms of constraint violation. An invariant-based specification is employed to describe the adaptation strategy. It did not address when multiple invariants are violated, how to cope with the potential underlying conflicts, or dependency between the underlying reconfiguration strategies. Later in [45], Cheng *et al.* took more than one adaptation dimension into consideration and provided a utility theory based approach for the strategy selection. However, the selection process was carried out at the application level, it still did not answer the question when multiple invariants were violated, whether the selected adaptation strategies would be conflicting or not.

Zhang and Cheng proposed a model-driven adaptive program development process [192]. By using petri nets and the state machine, the application behavior and dynamic adaption are analyzed in the model level. The notion of *overlap adaptation* deals with the coordinations among parallel adaptations. This approach focuses on the behavior modeling and change analysis, while our approach focuses on the structural changes.

Le Métayer demonstrated the feasibility of using conditional graph rewriting rules as a *coordinator* to describe dynamic evolution [115]. Wermelinger and Luiz Fiadeiro further investigated the concept and proposed an algebraic graph transformation based approach (double pushout) for architectural reconfiguration modeling [184]. But neither work included the conflict analysis in case of multiple reconfigurations.

In [90], Hausmann *et al.* presented a critical pair analysis based approach for the detection of conflicting functional requirements. The theory from graph transformation makes the conflicts and dependence relations of use case models more precise and accurate. It demonstrated the applicability of graph transformation in UML use case models. Mens *et al.* further widened the application domain of graph transformation, *e.g.*, refactoring dependence and conflict analysis [129, 130], and model inconsistency detection [131]. The methodology is similar to ours. They used the graph transformation rules to specify refactoring or model inconsistency detection/resolution. AGG was also employed as the underlying tool.

7.6 Summary

For architecture-based self-adaptive software systems, multiple reconfigurations are possible to be invoked due to various context changes. Ad hoc solutions are not satisfactory to address the potential conflicts and dependencies between the involved reconfigurations. A rigorous modeling and analysis is necessary. Throughout this chapter, we present a graph transformation based approach. An algorithm to map the transformation to the basic architectural reconfiguration command list is presented. We have also developed a tool VIDE which integrates AGG as an external utility to help with the architecture and reconfiguration rules design. Possibilities for future research include applying the approach to more practical and complicated applications, and tailoring the critical pair analysis algorithm to reduce the duplicate conflict or dependency report so as to better fit the problem context.

Chapter 8

Model Based Verification of Dynamic Evolution

CONTENTS

*This chapter is modified and updated from SCIENCE CHINA: Information Sciences 59(17) 2016:032101, with permission from Springer Science+Business Media.

As mentioned in previous chapters, self-adaptation is highly desirable for service oriented systems in an open environment. The dynamic evolution is a crucial intermediate enactment phase of the adaptation cycle. For the adaptation to be trusted, it is necessary to keep the dynamic evolution process consistent with the specification. We observe that hierarchical architecture styles are common in the service oriented system. As result, in this chapter, we present two kinds of evolution scenarios and propose a novel verification approach based on hierarchical timed automata to model check the underlying consistency with the specification. It examines the procedures before, during, and after the evolution process, respectively and can support the direct modeling of temporal aspects, as well as, the hierarchical decomposition of software structures. Probabilities are introduced to model the uncertainty characterized in open environments, and, thus, can support the verification of the parameter-level evolution. We present a flattening algorithm to facilitate automated verification using the mainstream timed automata based model checker – UPPAAL (integrated with UPPAAL-SMC). We also provide a motivating example with performance evaluation that complements the discussion and demonstrates the feasibility of our approach.

8.1   Introduction

With the development of network technology, the operating environment of modern software is not as isolated or closed as before. Such an environment, characterized by the openness and dynamism, inevitably affects the behaviors of the inhabitant systems, for example, the inherently uncertain nature of the network, the volatile user requirements, as well as, the quality of services. This ongoing trend has a significant implication on the software construction and calls for more flexible adaptability of software artifacts. Service oriented systems are becoming a promising paradigm for software in the age of the Internet. Since the ultimate goal of the software process is to deliver satisfying products without compromising the quality of service, the need for supporting dynamic adaptation is becoming an essential concern, especially in some mission-critical domains.

Dynamic evolution denotes the process of changing behaviors according to the contextual information so as to provide continuous and desired quality of services. By behavior, we mean the collection of software execution sequences, as well as, the states and their transitions of software systems. In [182], Wang *et al.* defined *dynamic evolution* as a specific kind of evolution that updates running programs without interrupting their execution. Thus, the main purpose is to minimize the cost of service unavailability or to avoid service degradation. For the process to be trusted, consistency with the specification is a prerequisite. As mentioned, the software artifacts are usually composed of a set of autonomous services in open environments, and the uncertainty from such environments becomes the driving forces of the evolution. The component services can be dynamically integrated or detached from the system, and, thus, the evolution is closely related to the architecture reconfigurations.

Having observed this, many efforts investigate the problem from an architectural perspective, using architecture description languages to model the system structure and the dynamism [137]. There indeed exists a causal relationship between software architecture and the changing environments (including the surroundings and requirements, etc.), but such a relationship is implicit. From end users' view, they care more about the running system's behaviors rather than the structure changes. Therefore, a modeling mechanism that can directly capture behaviors is more favorable. Secondly, in open environments, software components are composed to form services (*e.g.*, service composition), and this composition procedure is recursive. As a result, such systems naturally exhibit a hierarchical structure [119]. Moreover, time related properties are particularly relevant with the quality of the composed service, and, in many cases, violation of time constraints triggers the evolution process [105]. Timed automata have proven to be tremendously useful for the verification of time related properties over the decades with good tool support [5]. Moreover, the transitions of timed automata can also be extended with probability distributions, and this can be used to model the underlying uncertainty of involved systems [33]. However, the lack of high-level composable patterns for a hierarchical architecture design hampers its further application. Users have to manually cast those terms into a set of clock variables with carefully calculated clock constraints, and the process is tedious and error prone as well.

The above considerations motivate us to propose a novel approach based on hierarchical timed automata to model and analyze dynamic evolution process during self-adaptation. Compared with the existing work, our approach makes the following contributions: 1) the approach can directly support the modeling of temporal concerns, hierarchical structure, state transitions, and evolution behaviors; 2) we incorporate features of probabilistic transitions to model the uncertainties during operations characterized in open environments. By using the statistical model checking tools, our approach can support the verification of probability related properties; 3) we use both classical temporal logics as well as the probabilistic temporal logics to specify the properties that are supposed to be held, and these properties can be verified before, during, and after the dynamic evolution, respectively; 4) to render the model amenable to analysis by existing tool-sets, we also propose a translation algorithm to flatten the hierarchy. To better describe the verification technique, we further revise and enhance the illustrative example taken from the online business domain as widely adopted in the literature [164, 191, 196] with more features—such as probabilistic transitions—and a more comprehensive set of experiments by scaling up the size. Moreover, a more rich set of properties to be verified has been presented to demonstrate the feasibility of our approach.

The rest of this chapter is organized as follows. Section 8.2 introduces some theoretical background and the motivating example; Section 8.3 illustrates our approach; Section 8.4 presents the flattening algorithm and verification with performance evaluation; Section 8.5 discusses our approach, and Section 8.6 compares our work. Section 8.7 concludes this chapter.

8.2 Background

Timed automata are an extension of finite state automata to model real-time systems. Given the introduction of *clock*s, a run of the automaton along a sequence of consecutive transitions is of the form: $(l_0, v_0) \rightarrow (l_1, v_1)... \rightarrow (l_p, v_p)$, where $(l_i)_{0 \le i \le p}$ denotes the location, $(v_i)_{0 \le i \le p}$ denotes the clock value, and (l_i, v_i) constitutes the state. Specifically, l_i is associated with some combinations of clock-related Boolean expressions to denote time invariants, marked as $Inv_{l_i}(v)$. Since time is continuous, timed automata theoretically have infinite many states. By techniques such as region construction, these infinite states can be reduced to finite states [5], and an interesting property of timed automata is that *reachability* is decidable.

Hierarchical timed automata (HTA) introduce a refinement function to describe the hierarchy relationship between states. Therefore, they can naturally model the complex structures, for example, the composite states with several regions. Given the fact that in open environments, systems are usually constructed through sets of atom/composite services, the HTA intuitively corresponds to such complex hierarchical structures. To facilitate discussion, we have given related definitions of HTA formally, and then we will use an online order processing system as the motivating example to illustrate our approach.

Definition 8.1   Timed automata are defined as a tuple $\langle S, S_0, \sigma, C, Inv, M, T \rangle$, where

■ S is a finite set of locations,

■ S_0 is the finite set of initial locations,

■ $\sigma : S \rightarrow \{BASIC, COMPOSITE, INIT\}$ is a type function, that is, σ returns a specific type of a location,

■ C is a finite set of clocks,

■ M is the set of events, which are two types: synchronized (coordinated by ! and ?) and asynchronized, and

■ $T \subseteq S \times (M \times CC \times 2^C) \times S$ is a finite set of transition steps in which CC is a finite set of clock constraints, and 2^C denotes the power set of reset clocks.

For example, if an action a happens and the valuation of clocks satisfies the constraint g, it can cause a location s at time t to change to s'. This transition *trans* can be formalized as $\langle s, m, g, r, s' \rangle$ in which $m \in M$, and any clock, which is included within r, is reset and restarts from 0 as soon as the transition happens. For clarity, sometimes the transition step can also be written as $s \xrightarrow{m,g,r} s'$. For each transition $t \in T$, there are two associated functions, mapping to the related source and target locations, respectively, *i.e.*, $SRC : T \to S$, and $TGT : T \to S$.

Definition 8.2   Hierarchical timed automaton can be defined as a tuple $\langle F, E, \rho \rangle$, where

- F is a finite set of timed automata,

- E is the finite set of events, and

- ρ is a refinement function, mapping a location to a set of automata, *i.e.*, $\rho : \bigcup_{A \in F} S_A \mapsto 2^F$ in which S_A denotes the location set of each element automaton in F. ρ constructs a tree (hierarchical) structure to the related automata.

Based on the refinement function, we can *recursively* define ρ^* to denote the set of descendent timed automata of a composite location s: $\rho^*(s) = \rho(s) \cup (\bigcup_{s_i \in S_{\rho^*(s)}} \rho(s_i))$. There is a special subset of descendent timed automata of a composite location s. For each element of such subsets, denoted by $\rho_{leaf}(s)$, there are no further children timed automata, *i.e.*, $\rho_{leaf}(s) = \{A' | A' \in \rho^*(s) \wedge \forall s' \in A', \rho(s') = \emptyset\}$. If a transition t is involved with composite states, either the source or the target is composed of several orthogonal basic states. To model this, we utilize source and target restriction functions, which are defined as follows, $sr : t \to S$ where $S \subseteq \bigcup_{A_i \in \rho^*(SRC(t))} S_{A_i}$ and $tr : t \to T$ where $T \subseteq \bigcup_{A_i \in \rho^*(TGT(t))} S_{A_i}$. For these two sets, elements are pairwise orthogonal locations, if any. As their names suggest, these functions are mainly to restrict the transitions involved with those composite states, and they are yielding the actual sources or targets of a transition. Because of the composite location, it is possible that there are multiple active locations at the same time, we define the set of all the active locations as a *configuration*, and those active locations extended with the values of their corresponding clocks could be similarly defined as *timed configuration*. It can be regarded as a snapshot of the system.

In many cases, the software systems suffer from the dynamism and uncertainty in the environment [119], such as the Internet, and exhibit several kinds of random behaviors to certain degree. To enable the modeling of the underlying uncertainties, we extend our notion of timed automata and hierarchical timed automata with the ability to allow the probabilistic transitions as defined in the following.

Definition 8.3   **Extended timed automaton** can be defined as a tuple $\langle S, S_0, \sigma, C, Inv, M, Prob \rangle$, where

- S, S_0, σ, σ, C, and M are the same as those defined in the previous *timed automaton*, and

- *Prob* $\subseteq S \times M \times CC \times Dist(2^C \times S)$ is a finite set of probabilistic transition steps in which $Dist(2^C \times S)$ denotes the set of all probability distributions over $2^C \times S$, i.e., the set of all possible functions $\mu : (2^C \times S) \rightarrow [0,1]$ such that $\Sigma_{q \in 2^C \times S} \mu(q) = 1$. This probabilistic transition *pt* can be formalized as $\langle s, m, g, \mu, s' \rangle$. Similarly, for each probabilistic transition *pt* $\in$ *Prob*, there are two associated functions, mapping to the related source and target locations, respectively, i.e., $SRC : T \rightarrow S$ and $TGT : T \rightarrow S$.

Different from the probabilistic timed automaton given in [89], for simplicity, we do not allow the nondeterministic probabilistic transitions.

Definition 8.4   **Extended hierarchical timed automaton** can be similarly defined as a tuple $\langle eF, E, \rho \rangle$. The only difference from the previous notion of *hierarchical timed automaton* is that eF is the finite set of *extended timed automata*.

Statistical model checking is a very important approach of analyzing probabilistic models. Essentially it relies on finitely many runs of the stochastic model to get the samples, which further provides the statistical evidence for the satisfaction or violation of the specification based on hypothesis testing [116]. Different from the traditional model checking, it can be regarded as a trade-off between testing and verification and can express quantitative properties of interest. Also different from probabilistic model checking, it does not explore the whole state space, and, thus, can scale up to larger systems that may not be feasible for traditional numerical approaches.

8.3 Behavior Modeling

In the sequel, we first present an overview of our approach, and then we use a concrete example taken from the e-business domain to illustrate our approach. Online order management is one of the most important activities in e-commerce. It involves the collaboration of multiple services and also has strict temporal requirements. A complete e-commerce workflow typically consists of client-side service and vendor-side service. On the vendor side, it further consists of payment management service, goods management service, and delivery management service. To reduce unnecessary complexity without sacrificing the integrity, we only focus on the vendor side and simplify the client side service.

Generally, our approach consists of the following steps. First, we leverage

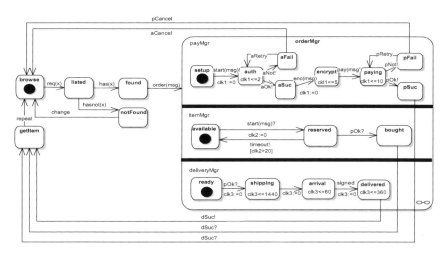

Figure 8.1: Example model illustration

hierarchical timed automata to construct the system models under design. To cope with the uncertainty aspects, we leverage the extended hierarchical timed automata to model the probabilistic transitions. Particularly, we model the system before, during, and after evolution, respectively. Second, we use temporal logics to specify the properties that are supposed to hold. Third, to verify against these properties, we use an indirect way to translate it to a set of sequential timed automata, *i.e.*, we flatten the hierarchy so as to make it verifiable via existing model checkers. In this section, we mainly discuss the behavior modeling step, and the verification related steps are explained in the next section.

Figure 8.1 illustrates the behavioral modeling of the example. Clients browse the web site and send requests. The results indicate two branches. If the vendor has the item, the clients will order and three concurrent threads will execute, *i.e.*, payment management (*payMgr*), item management (*itemMgr*), and delivery management (*deliveryMgr*). In payment management part, the process will start from *setup* state and authenticate the client's identity (*auth*). Normally, there is a time threshold for the authentication, and, in the example, we set it at 2 minutes. After passing the authentication, the following interactions will be encrypted, the time limit is set to be 5 minutes at most, and the online paying step lasts at most 10 minutes as set in the example. For the item management aspect, once the item is ordered, it will be transferred to the reserved location and kept in that location for at most 20 minutes. After that period and no payment message is received, it will be back to the *available* location again; otherwise, notified by the payment acknowledgement, it will be transferred to *bought* location. For the delivery management part, once the item has been paid for successfully, the initial *ready* location will be transited to *shipping*. The attendant temporal constraints states that the item should finish shipping within one day (1440 minutes). Once arriving at the destination, it will be dispatched at most 1 hour (60 minutes) and delivered to the clients within the next 6 hours (360 minutes).

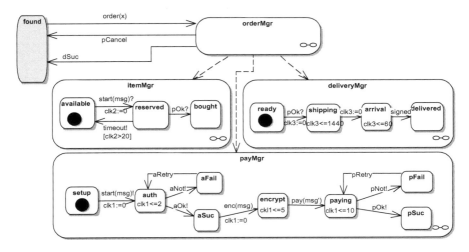

Figure 8.2: Hierarchy decomposition

Figure 8.2 illustrates the partial hierarchy decomposition of the example. Since *orderMgr* is a composite location, based on the definition of refinement function, $\rho(orderMgr) = \{payMgr, itemMgr, deliveryMgr\}$.

However, as discussed previously, there are various sources of uncertainty from the open environment. As Baresi *et al.* state, the world is open to new components that context changes could make dynamically available, and systems can discover and bind such components dynamically to the application while it's executing [15]. The service integrator might strengthen the security aspects, and want to add a new *sms* (short messages) authentication feature to the system on the fly. This requirement inevitably results in a dynamic evolution process. In *sms* authentication, it also has a strict time validity requirement, we can still use a clock variable *clk1* to model this concern. In the example, we set it to be no more than 8 minutes as time constraints ($clk1 <= 8$). The evolved behavioral model is partially described in Figure 8.3.

Figure 8.1 and Figure 8.3 describe the system models before (source) and after (target) the evolution, respectively. System states should be handled carefully to keep the evolution ongoing in a safe and low-disruptive way. Kramer *et al.* firstly proposed the notion of *quiescent state* as a sufficient condition [109] for a node to be safely manipulated from a running system. Later, Vandewoude *et al.* [178] proposed an alternative mechanism: *tranquility*. Given a state *S*, if it is not currently engaged in a transaction that it initiated, or it will not initiate new transactions, or it is not actively processing a request, or none of its adjacent nodes

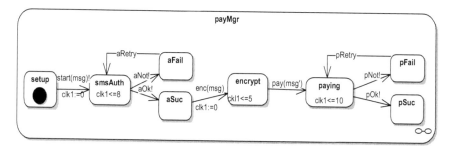

Figure 8.3: Evolved behavior model

are engaged in a transaction in which it has both already participated and might still participate in the future, we can predicate the state is in a *tranquil* status. Compared with the previous one, *tranquility* has weaker requirements and causes lower-level disruption. In our approach, we adopt *tranquility* as the premise for the evolution, but our approach is not bounded with any particular such mechanism. Given the definition of *tranquility*, we can impose the prerequisite for the dynamic evolution: the intersection of the active configuration set and evolution related element set is empty. For details on the *tranquility* mechanism, we refer interested readers to [178]. In the example, the evolution related service is an authentication service. So, if the active configuration set contains any element of the related set, such as *auth,aFail,aSuc*, it would be forbidden to enact the evolution; otherwise, it would be safe. As described in Figure 8.4, we take configuration $<< payMgr.setup >, < itemMgr.available >, < deliveryMgr.ready >$ as an example to explain the evolution model. The evolution action is marked as *evolve*, and the parameters are passed through messages.

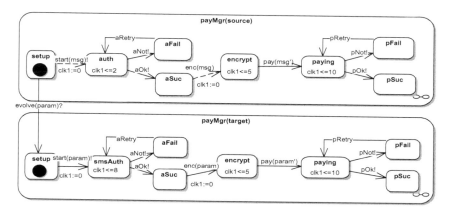

Figure 8.4: Behavioral model of evolution

Since the evolution action does not involve *itemMgr* location or *deliveryMgr* location, the tuple element $< itemMgr.available >$ and $< deliveryMgr.ready >$ will not be affected by the evolution. For clarity, Figure 8.4 only displays the evolution related parts. Tranquility requires the interactions between replaced nodes and the environments are blocked until the end of the substitution process. Therefore, the related transition *setup* $\xrightarrow{start(msg)!,clk1:=0}$ *auth* and *aSuc* $\xrightarrow{enc(msg),clk1:=0}$ *encrypt* are marked as dashed lines, representing *restricted*.

The second type of evolution case involves the updated model of *sms* service. Because of the uncontrollable nature of the underlying environment, it is quite possible that the message sent is lost during transmission. To characterize this, we use probability as the parameters of the quality of the *sms* service. Therefore, this kind of evolution mainly deals with parameter adaptation [62]. To illustrate this, we elaborate on the *smsAuth* state and make it a composite one, but it contains only one region. Figure 8.5 describes the activity. The inner behavior starts as soon as the process enters the composite state *smsAuth*. A short message is first sent (*smsSend*) to the customer within 2 minutes but without the guarantee of delivery due to the uncertainty of quality of service or the underlying environment. A probability ratio is assigned to the successful and failed receipts of the message, respectively based on preliminary knowledge or the experience from domain experts. In our example, we set the message loss (*lost*) possibility to be 1/5 and successful delivery to be 4/5 for default initial value. The diamond notation in Figure 8.5 represents the probabilistic choice. In case of message loss, it will repeat until being timeout (*timeOut*). Otherwise, it will enter the *received* state. From Figure 8.5, we can derive the refinement relation, i.e., $\rho(smsAuth) = \{smsSend, lost, timeOut, received\}$.

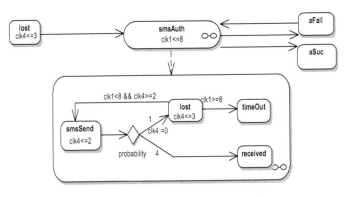

Figure 8.5: Hierarchy for smsAuth

8.4 Verification

In this section, we explain the verification of the evolution process based on the proposed models. The first kind of evolution involves the structural changes due to the replacement of existing components with different functional components. The verification has three stages, *i.e.*, before the evolution, during the evolution, and after the evolution. In each stage, we conducted the verification against the properties, and these properties are specified using computation tree logic (CTL). By *before evolution*, we mean the design artifact of the original system. By *after evolution*, we mean the updated models, which expose a different behavior sequence; while by *during evolution*, we mean the adaptation model connecting both the original and updated designs. As for the second kind of evolution, it mainly involves the parameter changes, aiming at non-functional properties. The model evolves without structural changes. Since the existing mainstream model checkers do not directly support the concept of hierarchy, to make it amenable for the direct analysis using available tool-sets, we first need to preprocess the model, including flattening the hierarchy and translating to a set of sequential timed automata.

8.4.1 Flattening algorithm

In our work, we use UPPAAL [20] (and its extension UPPAAL-SMC [50] for statistical model checking) as a test bed since its formalism is also based on timed automata, and this semantic similarity eases the verification process. As mentioned, we need to flatten the hierarchy before verification. In UPPAAL, each timed automaton is termed as the *template* and node as the *location*. We use UPPAAL's term in our translation and the flattening algorithm mainly has 5 steps. The pseudo-code is given in Algorithm 2.

1. For every timed automaton F in hierarchical timed automata M, which is not the root automaton, add a new *location*, and mark it as *inactive* (committed type).

2. For every transition t whose target is a composite state s, add the transitions from the *inactive* state to the default entry state inside s, the triggering event is the same as t's event. If the s itself is the entry state, the default entry state inside s is marked as entry state; otherwise, the state *inactive* is marked as the entry state.

3. To synchronise the execution of multiple related automata, the event of the transitions whose targets are composite states is extended to the broadcast channel and marked as send type (!); meanwhile, in its children timed

---

In UPPAAL, a committed location cannot delay and the next transition must involve an outgoing edge of at least one of the committed locations.

automata set, the transitions from *inactive* state to the default entry state are also extended to broad-cast channel and marked as receive type (?).

4. For every non-root timed automata, add a special state (marked as *committed*), and add transitions from each element of the source restriction set of the ongoing transitions from the composite state to this special state, and extend this transition's event to the broad-cast channel, and marked as receive type (?).

5. Add the transitions from each special state resulted from the previous step to the *inactive* of the first step in the same timed automaton.

Algorithm 2 describes the steps of the translation. It contains four sequential loops. Assume M is the HTA model, and n is the number of total locations within M. The cost of each cycle in the first loop (Line 3-7) is denoted by c_1. Then the cost of the first loop is $O(c_1 \times |M.F|)$. Since the theoretical maximum value of $|M.F|$ is n, we can deduce that the cost of the first loop is $O(c_1 \times n)$, *i.e.*, $O(n)$. Similarly, we can calculate that the cost for the third loop (Line 27-31) is also $O(n)$. As for the second loop (Line 9-26), it further has three nested sub-loops (Line 13-16, Line 11-22, and Line 10-25, respectively). We first consider the most inner sub-loop (Line 13-16), given a specific location s, the maximal number of transition targeted at is n, therefore, the corresponding cost is $O(n)$. The second inner sub-loop depends on the number of $|\rho(s)|$, and the maximal value is $n-1$. The outer sub-loop (Line 10-25) depends on the number of $|A_i.S|$, and the maximal value is $n-1$. So, the total cost for the second loop is $O(n^4)$. For the fourth loop (Line 32-43), since the maximal number of $|t|$ is n^2, where $t \in \bigcup_{A_i \in M.F} \Sigma_{A_i}$, we get the corresponding cost for the fourth loop to be $O(n^4)$. So, in total, the cost of Algorithm 2 is $O(n^4)$.

8.4.2   Correctness of translation

In [195], we have elaborated the formal semantics of hierarchical timed automata with multiform time extensions. Although in our context, we do not provide our model with features of multiform time, they are of the same nature, *i.e.*, physical time. Nevertheless, our model can be regarded as a special degenerated case of the model proposed in [195], thus, the defined semantics still hold. We also rely on the semantics of some model elements from UPPAAL. For example, *committed* type nodes. They are mainly related to the auxiliary locations introduced and are used only to enable or disable transitions. They do not take time, and, thus, are redundant for the configuration. We refer interested readers to [20] for details.

Based on the semantics, we use the notion of bi-simulation [132] to establish the behavioral equivalence between the hierarchical model and the set of flattened ones. For the two models to behave in the way that one simulates the other

	Data: M: Hierarchical timed automata models
	Result: T: Set of timed automata models
1	**begin**
2	$T \longleftarrow \emptyset$;
3	HashMap *map*;
4	**foreach** $A_i \in M.F$ **do**
5	create a template t; add locations and transitions based on $A_i.S$ and $A_i.\Sigma$;
6	**if** $A_i \neq A_{root}$ **then**
7	add $A_i\_inactive$ location in t; mark $A_i\_inactive$ as committed;
8	$T \longleftarrow T \cup \{t\}$; map.add($A_i$,t);
9	**foreach** $A_i \in M.F$ **do**
10	**foreach** $s \in A_i.S \wedge \sigma(s) = COMPOSITE$ **do**
11	**foreach** $A_i \in \rho(s)$ **do**
12	temp $\longleftarrow$ map.get(A_i);
13	**foreach** $in\_tr \in \{incoming\ transitions\ to\ s\}$ **do**
14	add transition *trans* from $A_i\_inactive$ to the entry location in temp;
15	augment the action of *trans* with *broadcast* channel;
16	**if** s *is initial* **then**
17	default entry location in A_i is marked as initial in temp;
18	**else**
19	$A_i\_inactive$ is marked as initial in temp;
20	map.update;
21	T.update;
22	**foreach** $A_i \in M.F \wedge A_i \neq A_{root}$ **do**
23	temp $\longleftarrow$ map.get(A_i);
24	add join location A_i_join in temp;
25	mark A_i_join as committed;
26	**foreach** $t \in \bigcup_{A_i \in M.F} \Sigma_{A_i}$ **do**
27	**foreach** $s \in sr(t) \wedge (\sigma(s) = COMPOSITE)$ **do**
28	**foreach** $A_j \in \rho^*(s)$ **do**
29	temp = map.get(A_j);
30	add transitions *tt* from each locations generated by $A_j.S$ to A_j_join in temp;
31	add transitions *tt'* from A_j_join to $A_j\_inactive$;
32	associate *tt* with channels, guards and clock resets based on t;
33	map.update;
34	$T \longleftarrow \bigcup_{A_i \in M.A}$map.get($A_i$);

Algorithm 2: Flattening algorithm

and vice versa, we have to demonstrate that there exists a binary relation R that both R and R^{-1} are simulations. Before proving the bi-similarity with respect to reachability, we first introduce the following lemma.

Lemma 8.1

Let TAs be translated from the HTA by Algorithm 2, a hierarchical state (s, u) in HTA is reachable, if and only if, the corresponding state (s', u) is reachable in TAs.

Proof 8.1   We first consider the sufficiency direction.

1 Assume (s, u) is the entry state in HTA, according to the translation algorithm, regardless of whether it is in the root automaton or not, the corresponding state after translation is (s', u). Since the inserted inactive location is marked as committed and the transitions will take place without delay and reach the default entry state in TAs. So, corresponding (s', u) is reachable in TAs.

2 Assume (s, u) is an intermediate non-entry state, it means there is at least one incoming transition. For such case, we use induction on the path length n to prove.

   2.1 When $n = 0$, this degenerates to the case of default entry state, and from the above statements, we know it holds.

   2.2 Assume when $n = k$ the conclusion holds, and there is a transition t targeting at (s, u), i.e., $(s_k, u_k) \xrightarrow{t} (s, u)$, which makes the path length equal to $k + 1$. According to the translation algorithm, if $\sigma(s) = BASIC$, the corresponding (s', u) is faithfully rendered by our algorithm and the same event of s can trigger the corresponding transition t' in the translated automata, i.e., $(s'_k, u_k) \xrightarrow{t'} (s', u)$. If $\sigma(s) = COMPOSITE$, according to the translation algorithm, we know that every composite state in the HTA model corresponds to several UPPAAL automata based on its refinement function ρ. But the transition t has been extended with the broadcast ability to synchronize these multiple automata and the corresponding state (s', u) can also be reached through the transition t', i.e., $(s'_k, u_k) \xrightarrow{t'} (s', u)$. In either possibility, the transition including the triggering events name, clock constraints are the same, and, thus, clock evaluations u carries over without changes. We can assert that there is a corresponding (s', u) reachable in the translated automata set (TAs).

For the necessity direction, we can analogously prove by induction on the length of path targeting at s and s_0 similar to the sufficiency direction introduced above.

Theorem 8.1

Let TAs be the translated set of UPPAAL *automata from the HTA by Algorithm 2, the two models are bisimilar.*

Proof 8.2 If we do not consider the notion of a *start* location, and the automata can be degenerated to a common labeled transition system (LTS). According to LEMMA 1, given every state (s, u) in HTA, we have a corresponding state (s', u) reachable in TAs, we can establish an equivalence relation R, in order that $((s, u), (s', u)) \in R$ and based on the definition of R, we can say that (s', u) is equivalent to (s, u). Thus, straightforwardly, an active configuration C in the HTA is equivalent to an active configuration C' in TAs if for all $(s, u) \in C$ there is an equivalent $(s', u) \in C'$ and vice versa. Since those auxiliary locations added by our algorithm transit to other meaningful states immediately and will not be part of any configurations, we can also establish a converse relation R^{-1}, such that $((s', u), (s, u)) \in R^{-1}$. Given the relations R, R^{-1} and LEMMA 1, we can conclude that the R is a bi-simulation of the derived LTS, and the two models are thus bisimilar.

Although Algorithm 2 targets the hierarchical timed automata, we can substitute that with the extended hierarchical timed automata, and then we get the translation algorithm for extended hierarchical timed automata. Since the only difference is in the probabilistic distribution over transitions, and we do not change the semantics of that during translation, thus, similarly we can prove the correctness for the extended hierarchical timed automata. Due to space limitations, we do not include that in our approach. For simplicity, we do not utilize the full expression power of UPPAAL-SMC, as it supports the priced timed automata. Instead, our extended timed automata can be regarded as a special case of priced timed automata in which the clocks evolve at the same rate.

8.4.3 Consistency verification

Once the hierarchical timed automata have been translated into a collection of concurrent timed automata, we can use existing model checkers to verify interested properties. As mentioned, we mainly leverage UPPAAL (integrated with UPPAAL-SMC, ver. 4.1.18 with academic licence) due to the similarities of the underlying models. For the first type of evolution, the verification is conducted in three stages, *i.e.*, before, during, and after the evolution. The four example properties are categorized into two types: one is safety, and the other is liveness. For the second type, we use statistical properties to specify user's requirement and quantitative model checking to verify whether or not the model satisfies such kinds of properties.

For the safety category, regardless of whether the system evolves or not, when the customer pays for the item, it should not be in the *available* state. Put it in a CTL formula, we get $P1 : A\Box(\neg(payMgr.paying \wedge itemMgr.available))$; the other interesting property is that once the customer pays for an item successfully, the item will be no longer in the *available* state and in the delivery management, neither will it stay in the *ready* state again. Then similarly, we get $P2 : A\Box(\neg(payMgr.pSuc \wedge (itemMgr.available \vee deliveryMgr.ready)))$. As to the liveness category, one property that the system should hold is: as long as a requested item is found, the item will be either *bought* or *available*, but not in the *reserved* state. In CTL formula, we get $P3 : A \diamond (found \to (itemMgr.bought \lor itemMgr.available))$; as another example of such kind, once the customer pays successfully, eventually, the item will be delivered. This property can be specified as follows: $P4 : A \diamond (orderMgr.pSuc \to deliveryMgr.delivered)$.

Based on the flattening algorithm in the previous section, we translate the system before the evolution into the four concurrent automata recognized by UP-PAAL as partially illustrated in Figure 8.6, *i.e.*, the sub-models of *itemMgr*(a), *deliveryMgr*(b), *payMr*(c), and *orderMgr*(d). Given the model and the specification, we verified the design and found that it satisfied all the above mentioned properties.

We continued to verify the system after evolution against the interested properties. The result is that the system satisfies the third and the fourth property but not the first or the second. By inspecting the counter example, we realized that the time limit of the new *sms* authentication module is too long. Since by combining the time spent on the encryption and the payment, it is possible that before the timeout of payment, the reservation of the item will expire and roll back to the *available* state. Therefore, it is possible that the customer has paid successfully, but the item's status is still available, leading to violation of $P1$ and $P2$ (*i.e.*, $payMgr.paying \land itemMgr.available$ and $payMgr.pSuc \land itemMgr.available$). Moreover, once the system is in such a configuration, since the channel of $payMgr.pSuc$ could not synchronize with the item, this will lead to a deadlock situation. In every individual step, the time requirement is fine per se, but when put together, inconsistency would happen. We investigated our original design given the traces generated by the counter example and found that changing the item's state to *reserved* before passing the authentication is not reasonable. Consequently, we revised our design and postponed the transition to the *reserved* state after passing the authentication. The reason why the first verification failed to disclose such a design flaw is that before evolution, the original authentication time plus the encryption time is still less than the predefined threshold. The revised design is given in Figure 8.7, which only includes the changed part. The transition from *aSuc* to *encrypt* has been renamed to *start(msg)!*, and, thus, the transition from *setup* to *smsAuth* has been renamed. We again model checked the revised design against the four properties and this time, all of them were satisfied.

The last step is to verify the evolution phase. Since we revised the model as described in Figure 8.7, the updated behavioral models are instead translated to the UPPAAL's input templates. Again, we model checked the design against the four properties above and concluded that all of them hold. That is, the configuration, including $<< payMgr.setup >, < itemMgr.available >,$ $< deliveryMgr.ready >>$, behaves consistently with the required safety and liveness properties. Figure 8.8 illustrates the UPPAAL models for the evolution phase. Since the tranquility mechanism requires that the authentication services

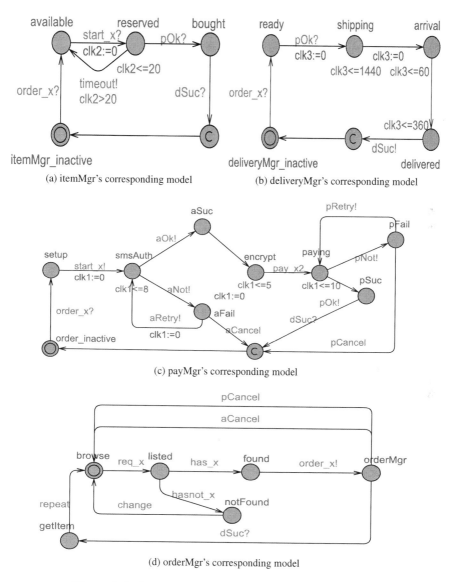

(a) itemMgr's corresponding model

(b) deliveryMgr's corresponding model

(c) payMgr's corresponding model

(d) orderMgr's corresponding model

Figure 8.6: Corresponding UPPAAL models illustration

should be inactive so as to be replaced, the related nodes and transitions are deactivated accordingly. Service information before evolution, such as user data, is transmitted to the target nodes as messages.

To scale up the size of our example, we use the number of repository items as the parameter (denoted by N). We measured the number of generated states and the consumption of memory needed to perform the complete verification.

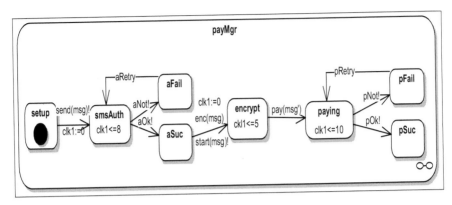

Figure 8.7: Revised design of payMgr

The experiments are conducted on a PC with an Intel i7-3770 3.4GHz processor and 4.0GB RAM running 64-bit Windows 7 Professional Operating System. The versions of UPPAAL (*verifyta.exe*) and jdk are 4.1.18 with academic licence and 1.8.0, respectively. The complete experiment data (*i.e.*, before, during, and after the evolution) gathered are given in Table 8.1 and Table 8.2. Besides, after revising our original design, we reconducted the experiments and these data results are also given in these two tables.

Table 8.1: Summary of the experiment results[†]

N	Before evolution		During evolution		After evolution	
	States	R/V (MB)	States	R/V (MB)	States	R/V (MB)
1: P1	1158	7.4/30.5	22	3.2/21.2	29	7.4/30.5
1: P2	1230	7.4/30.5	24	3.6/22.4	41	7.5/30.5
1: P3	1270	7.6/30.9	1828	3.2/29.5	1398	7.6/31.0
1: P4	1294	7.6/31.1	2034	3.5/29.0	1454	7.7/31.1
6: P1	182646	17.1/49.6	27	7.7/55.6	65	7.5/30.6
6: P2	182974	17.3/50.0	34	8.2/53.3	206	7.6/31.0
6: P3	183634	17.5/50.4	642268	8.9/61.8	626962	41.4/96.6
6: P4	249054	18.4/52.0	1316274	8.6/62.0	956902	46.0/105.9
10: P1	10496886	573.0/1146.7	31	7.6/30.8	111	7.5/30.7
10: P2	10497214	574.6/1149.8	42	7.8/31.3	508	7.8/31.2
10: P3	10507474	576.3/1153.2	NC	OM	NC	OM
10: P4	12070622	585.4/1171.5	NC	OM	NC	OM

[†] N = number of items, R/V = Residential/Virtual Memory, NC = Not Concluded, OM = Out of Memory.

For Properties *P1* and *P2* in the original design, since they do not hold, and

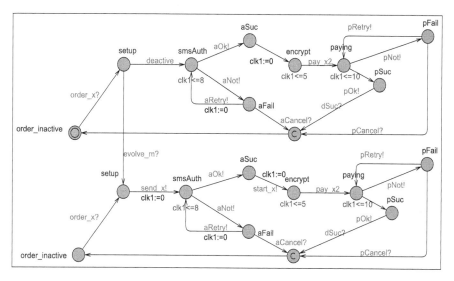

Figure 8.8: Evolution model illustration

Table 8.2: Summary of the experiment results (revised model)[†]

N	Before evolution		During evolution		After evolution	
		R/V	(Revised)	R/V	(Revised)	R/V
	States	(MB)	States	(MB)	States	(MB)
1: P1	1158	7.4/30.5	2952	7.5/30.6	1152	7.4/30,5
1: P2	1230	7.4/30.5	3648	7.6/30.9	1220	7.5/30.6
1: P3	1270	7.6/30.9	3810	7.8/31.4	1254	7.6/31.0
1: P4	1294	7.6/31.1	4032	8.0/31.5	1284	7.7/31.1
6: P1	182646	17.1/49.6	391320	28.7/71.9	155440	15.9/47.0
6: P2	182974	17.3/50.0	467280	28.9/72.2	155764	16.0/47.3
6: P3	183634	17.5/50.4	469922	29.3/73.0	156294	16.3/47.8
6: P4	249054	18.4/52.0	732496	31.1/76.6	221844	17.0/49.4
10: P1	10496886	573.0/1146.7	25626960	1252.5/2519.2	8542000	484.2/969.1
10: P2	10497214	574.6/1149.8	25626960	1256.1/2526.3	8542324	485.9/972.6
10: P3	10507474	576.3/1153.2	25668002	1259.0/2532.1	8550534	487.1/974.8
10: P4	12070622	585.4/1171.5	31921488	1284.5/2583.6	10115732	497.0/994.9

[†] N, R/V, NC, OM are the same as those in Table 8.1.

UPPAAL can find a counter example without exploring the whole state space, the performance data collected (*i.e.*, states, memory consumption, and time) in Table 8.1 are only those explored before finding the counter example. We also conducted experiments on the evolution model with the original design, since the properties in our case study are global ones. Similarly, the same properties do not hold and the results are only those explored before finding the counter example. Since we only revised the evolution design, the models before evolution will not be affected. After revision of the evolved model, all the four properties now hold. As the size goes up, we find that soon the generated state space grows beyond the machine's capability, and we use *OM* (out of memory) and *NC* (not concluded) in the table to describe this phenomenon. Based on our analysis in Section 8.4.1, for our translation algorithm, the time cost is still polynomial to the number of locations (n) in HTA, *i.e.*, $O(n^4)$, and the maximal number of auxiliary nodes is linear to n. Thus, the main reason for the state explosion is introduced by the parallel composition of timed automata. In Table 8.1 and Table 8.2, we can find that even for the same start system, the number of states explored is usually different for different properties. This is due to the optimization strategies utilized by UPPAAL. Generally, during the evolution, since it involves two versions of the models, the number of states are much more than the other two stages. This can be observed in Table 8.1 and Table 8.2. Theoretically, timed automaton has an infinite number of states because of the dense representation of time. By using the notions of region reduction, the infinite becomes finite, making the state exploration possible for model checking [5]. For details of state space generation and exploration techniques in UPPAAL, please refer to [20].

As to the second type of evolution, users might be interested that in spite of the uncertainty of message loss, given a period of time (e.g., 8 time minutes in *smsAuth*), the probability of successful delivery of the message should be above a certain threshold, for example, more than 96%. Following UPPAAL's syntax for probabilistic temporal logics, we get *P5*: $Pr[< 8](\diamond smsAuth.recived) >= 96\%$. Based on the composition hierarchy given in Figure 8.5 and the translation algorithm, the resulted flattened UPPAAL models are described by Figure 8.9. To make it workable for UPPAAL SMC, we still need to moderately revise the original payMgr model. As can be observed in Figure 8.9(a), the location type of *aSuc*, *aFail*, *pSuc*, and *pFail* has been changed to the *committed* type since it will transit instantaneously without time cost; while in Figure 8.9(b), those locations with time invariants will enable the transition stochastically. Particularly, the transitions take place according to uniform distribution [50]. Therefore, we could use the given parameters of the model (design-time estimates) to conduct the statistical verification.

The curve in Figure 8.10(a) illustrates the cumulative probability distribution for the successful delivery of a message as time goes by based on the initial parameters. Given the initial estimate for the successful message delivery being 0.8, it is clear that the property *P5* is satisfied. However, as argued in [62], the estimates are seldom correct. The problem is that these actual parameters may change over time. In the running phase, real-time values are needed to update the estimates in order to make the model faithfully reflect real behaviors and thus be useful for analysis or prediction. If we have external gauges in the implementation of the system recording the message loss times and success delivery times, respectively, we could update the model parameters and thus verify whether the model continues satisfying the requirement more accurately. We use simulations to illustrate this. For example, after a certain period of running, the data collected tell that the initial value for message loss are underestimated, and the actual value should be up to 1/3. Based on this new ratio value, the model checker finds that *P5* does not hold any longer. Figure 8.10(b) illustrates the updated cumulative probability distribution for successful message delivery over time. The cumulative probability given the time limit is only approximately 91%. Once the violation of the requirement is detected, the system could evolve by replacing the component service, *i.e.*, *smsAuth* in our example, with one of higher reliability for message delivery until the property is satisfied.

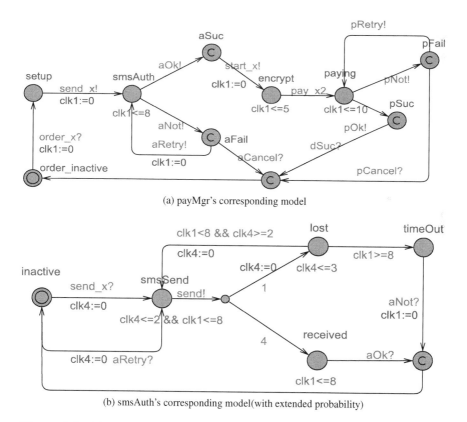

(a) payMgr's corresponding model

(b) smsAuth's corresponding model(with extended probability)

Figure 8.9: Flattened UPPAAL models for the two synchronized automata: payMgr and smsAuth

For the experiments to verify Property *P5*, we set the values of false negative (α) probability, false positive probability (β), and uncertainty probability to be all 0.001. Thus, it can give a relatively accurate result with the confidence rate to be 0.999.

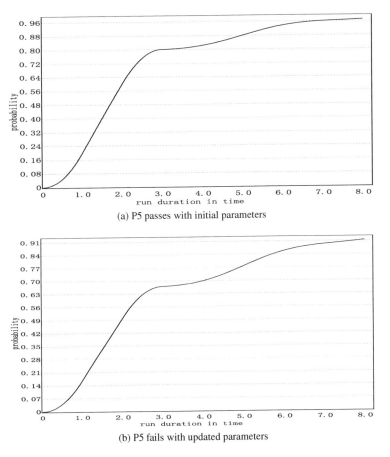

(a) P5 passes with initial parameters

(b) P5 fails with updated parameters

Figure 8.10: Cumulative probability for successful message delivery given different parameters

8.5   Discussion

As described in Section 8.4, the states in our approach do not mean fine-grained, algorithm-level data values. Instead, they denote the coarse-grained, interface-level, observable information. This is due to the characteristics of open environments. Indeed, in such environments, the system integrators are not necessarily the component developers, and details of intra-component level states are hidden from the end-users. We also employ the concept of *tranquility* as the basis for the evolution. However, *tranquility* is not free and requires some assumptions from the environments. Although in our example, this is trivial to the verification result, since the timing of *itemMgr*'s *reserved* state is triggered after passing the authentication. Statistics indicate that only in very rare situations, the tranquility conditions are not reachable. Detailed explanations are presented in [178].

Hierarchical timed automata actually formalize a tree-like structure of the modeled system. It's partially motivated by the UML state machine diagram [134], in which the composite state imposes the hierarchy. However, involving composite locations, there are some transitions that directly point to or start from arbitrarily inner locations in composite ones. In UML superstructure specification, it states that 'if a transition terminates on an enclosing state and the enclosed regions do not have an initial pseudo-state, the interpretation of this situation is a semantic variation point. In some interpretations, this is considered an ill-formed model' [134]. Therefore, in our model, to avoid potential ambiguity, we do not consider such transitions. The approach holds an implicit assumption that normally such inter-level transitions will not take place between sub-locations in different composite ones, but of course, they can be synchronized through channels. That is the potential limitation of our approach.

In the motivating example, messages pass between different states to transmit related information, for example, the item number and user identity. However, in UPPAAL, it does not support the message passing. We use an indirect way to get around the problem; since, the number of messages is finite, and we encode these messages to different variables. During the transition, we manipulate the values of these variables and these values exactly correspond to the predefined messages. Verification is a broad issue, in this chapter we emphasize the consistency aspect. Thus, we did not address the service substitution assurance mechanisms [37]. In our approach, it is probable that there are multiple evolution strategies, each of which has a different set of source states, target states, and tranquility states. Despite the complexity it introduces to the analysis, on one hand, we gain a high level of flexibility; on the other, we can inspect the state transition traces of the system, which cannot be realized through traditional structure based models. In our current work of the extended timed automata, since we rely on UPPAAL-SMC for verification, it supports a limited number of stochastic models. For bounded delay in locations with invariants, uniform distribution is assumed; while for unbounded delay, exponential distributions with user-defined rates are used [50]. That's also the potential limitation of our work.

8.6 Related Work

Consistency assurance mechanisms of dynamic evolution are intensively studied in the literature. In [98], Huang *et al.* exploited the reflection mechanism to establish the causal relationship between the requirement/environment and the software artifact. In [121], Ma *et al.* leveraged inheritance and polymorphism to enact dynamic evolution. These works do not support formal consistency analysis. The works on model-based verification include [17, 39, 91, 192]. In [17], Baresi *et al.* studied the domain-specific features of publish-subscribe style-compliant software architectural styles and proposed customized state-generation algorithms, which effectively reduced the state space. Their work mainly verified whether the specific software architecture satisfies the design-level requirements; while ours mainly concentrates on the behavioral consistency during evolution. In [91], Hayden *et al.* proposed a merge-based approach for enabling dynamic evolution. Two versions of the program, *i.e.*, the old and the evolved, are bound and specified. Therefore, the consistency can be checked against based on the specification; a similar work includes [39]. They rely on specific language and corresponding compilers. This assumption usually does not hold in open environments. In [192], Zhang *et al.* tried to use Petri-Net to model the components' behavior and LTL formulas to model specification, thus utilizing existing tools to model check the interested properties. However, the approach neither addressed the temporal aspects of the system nor the hierarchical structures.

In [193], Zhang *et al.* proposed a tool chain to support the modeling of behaviors by the property sequence chart. The tools complement the UML sequence diagrams with rigorous semantics and feature the modeling ability of time and probability for service compositions. However, the work emphasizes extending the traditional scenario-based modeling language to model service compositions and does not address the dynamic evolution aspect. In [49], David *et al.* leveraged the expressive power of probabilistic timed automata to model the cyberphysical systems under varying environmental settings. In [185], Xu *et al.* studied the problem of detecting context inconsistency for Internetware computing. The proposed approach could efficiently detect contextual noises. However, these works do not address the evolution concerns either. In [33], Calinescu *et al.* argued the importance of applying quantitative verification techniques for dynamic evolvable systems in open environments. In line with this, they proposed a framework for the development of adaptive service-based systems [34]. In [65, 67, 69], Filieri *et al.* proposed a formal verification approach for self-adaptive software based on probabilistic model checking. Different from the approach here, these works mainly address the evolution of non-functional requirements related models, such reliability, quality of service, without considering the structural changes of software architecture models.

There have also been other proposals based on graph rewriting techniques (e.g., [96, 186]). In [96], Hölscher *et al.* presented semantics for UML based on the translation of a given model into a graph transformation system. The graph transformation system comprises transformation rules and a working graph representing current system states. These states are limited to simple states instead of composite ones as supported in our approach. Xu *et al.* [186] used attributed graph grammar to model the system and graph transformation to model the dynamic evolution. Graph based notation has the visual intuitiveness advantage. However, when the model scales up, the cost of verification grows quickly, due to the fact that the matching is implemented through graph morphism. Besides, all the above mentioned approaches of such category lack direct support for the modeling and analysis of dynamic evolution.

8.7 Summary

In this chapter, we discuss a hierarchical timed automata based approach to model and analyze the dynamic software evolution from a behavioral perspective. Specifically, our work can support the analysis of both functional evolution (with structural changes) and non-functional evolution (with parameter changes). To the best of our knowledge, this is the first approach employing hierarchical timed automata to model evolution process through a behavioral perspective. A motivating example with a comprehensive set of performance evaluations is discussed to illustrate the feasibility of our approach. In our approach, we used the notion *tranquility* as the basis for the dynamic evolution. However, the approach is not bound by the mechanism and is open to other alternatives, for example [16]. Despite the fact that our approach mainly targets the design phase, it can be also applied in the running phase; since, the boundary between different phases is now blurring. For example, the model can be kept alive and verified during runtime. This is particularly important for the parameter level evolution, since the actual data to update the estimate can only be collected at runtime. The readers are advised to consult to Chapter 9 in this regard.

Chapter 9

An Iterative Decision-Making Scheme for Adaptation

CONTENTS

*This chapter is based on Fundamental Approaches to Software Engineering, volume 9633 of the series Lecture Notes in Computer Science, pp. 269-286 (DOI:10.1007/978-3-662-49665-7_16), with permission from Springer Science+Business Media.

9.1 Introduction

The aim of this chapter is to introduce a formal method to support software adaptation, based on Markov Decision Processes . As mentioned in previous chapters, with the rapid development and wide application of computing and networking technology, the inhabitant environment of software is evolving to be more open and dynamic. Due to the new characteristics introduced by such an environment, software systems must be more autonomous, more sensitive to contextual changes, more reactive, and more adaptable. Software adaptation is usually governed by adaptation logic. It is advantageous to gather the complex adaptation logic into a separated component for at least three reasons: first of all, the occurrence of runtime phenomena is asynchronous with respect to the flow of the application logic; second, not all information about the phenomena is available at the design time; and third, the specification of the adaptive behavior may evolve over time.

Designing adaptation logic plays a central role in developing self-adaptive systems, which can be, in a large sense, cast into a classical *decision-making* problem. This is particularly suitable for a closed environment, where a decision model, such as a Markov Decision Process (MDP) [146], can be constructed before the running of the system. Optimal adaptive strategies can be determined by solving the MDPs. However, in open environments, the adaptation logic of self-adaptive systems is governed by empirical data which should be acquired at runtime and is subject to practical constraints. This process is much different from traditional decision problems, where a decision-making model is determined for reasoning. In many situations, one has to sacrifice the optimality of an adaptive solution to a certain extent in order to satisfy various Quality-of-Service (QoS) constraints.

An example is a Web system that provides news content services [169]. Suppose that the system detects high latency of content delivery at a certain moment, and that the system can lower the content fidelity (such as delivering multimedia contents in the text mode) and/or increase the server pool size, and that the benefits or costs of these operations are measured quantitatively. Furthermore, to achieve more sophisticated effects, operations can be combined to form a strategy. For example, one simple strategy could be: Once "high latency" is detected, increase the number of Virtual Machine (VM) instances by one; if "high latency" persists, switch from the multimedia mode to the text mode. Here, while the first detection of "high latency" is a precondition, its second (attempted) detection is related to the uncertainty of system operations. Because multiple strategies built into the adaptation logic may be triggered by the same condition, an additional mechanism is required to select one from them.

A key challenge of the strategy selection for the Web system is that some parameters of the underlying decision model, such as successful chances of operations, are not fixed. For example, if the VM number is increased by one, the probability that latency will drop below the threshold may increase, but it still has to be estimated based on runtime data. While the idealized goal is to select an optimal strategy, it is important to take into account the practical constraints. For example, obsolete data no longer reflects the current environmental situation; the time frame of data sampling may be constrained by the tolerance of adaptation delay; the sampling frequency may be restricted because of its performance overhead on the network; and last but not least, the adaptation should not downgrade the functional performance of the system by consuming too much computational capacity (*e.g.*, CPU and RAM). In short, besides decision accuracy, runtime decision-making has to address the limitation of data and computation resource.

The adaptation model for the Web system can be formalized as an MDP in which actions represent operations and schedulers represent strategies. The runtime data are stored as, *e.g.*, a set of matrices for estimating the transition probabilities of the MDP. Therefore, the problem of strategy selection reduces to the problem of finding a scheduler which can minimize the (expected) cumulative cost in an MDP with *empirically determined* transition probabilities and a given subset of schedulers. Despite this problem is well understood in the theory of MDPs [146], Su *et al.* [169] put forward an *Iterative Decision-Making Scheme* (IDMS) that supports a trade-off between accuracy, data usage, and computational overhead. The basic ideas of IDMS are as follows:

- both point and interval estimates of transition probabilities for the MDP decision model are carried out based on the data structure for runtime data;

- as the next step, a scheduler that minimizes the cumulative cost for a given reachability problem can be computed; and

- one then determines whether this scheduler meets a criterion called *confident optimality*. If yes, or if the maximal number of iterative steps is reached, the iteration terminates; otherwise, the iteration returns to data sampling.

In [169], three metrics for IDMS are formalized. These include:

1. the probability that a confidently optimal scheduler is truly optimal, namely accuracy;

2. the average sample size of the iteration, which is a direct metric of data usage; and

3. the average time of iteration, which measures computational overhead conveniently.

The tradeoff among these three metrics is realized by adjusting the criterion of confident optimality and the sample size during the iteration. The core method of IDMS is a value-iteration algorithm developed from probabilistic model checking [70].

Several high-level frameworks and approaches based on probabilistic model checking have been proposed to aid the design of self-adaptive systems, but with emphasis on different aspects of the adaptation [32, 34, 68, 80, 133]. In existing self-adaptation frameworks, operation effects [45] as well as other uncertain aspects (such as model drift [59], reliability prediction [46], and resource availability [144]) are characterized probabilistically. However, none of these works address the problem of making the aforementioned tradeoff in the adaptation.

IDMS can be naturally embedded into the Rainbow framework [45] which employs a standard, point-valued MDP as its decision model, and thus extends the adaptation function of the latter. The flexibility of IDMS can be demonstrated by a case study on a Rainbow system.

9.2 An Iterative Decision-Making Scheme

In this section, we present main stages and techniques of IDMS and describe the realization of trade-offs between the three metrics. However, let's start with some preliminaries, mostly on MDPs and their solution methods.

9.2.1 MDPs and value-iteration method

Definition 1 (MDP) *A Markov decision process is a tuple* $\mathcal{M} = (S, Act, \mathcal{P}, \alpha, C)$ *where*

- *S is a finite, non-empty state space,*

- *Act is a finite non-empty set of actions,*

- α *is the initial distribution over S,*

- $\mathcal{P} = \{\mathcal{P}_a\}_{a \in Act}$ *is a family of transition probability matrices indexed by* $a \in Act$, *and*

- $C : S \to \mathbf{R}_{\geq 0}$ *is a cost function.*

We require that, for each $a \in Act$ *and* $s \in S$, $\mathcal{P}_a[s,t] \geq 0$ *for all* $t \in S$ *and* $\sum_{t \in S} \mathcal{P}_a[s,t] \in \{0,1\}$. *We say action a is* enabled *at s if* $\sum_{t \in S} \mathcal{P}_a[s,t] = 1$.

Schedulers play a crucial role in the analysis of MDPs. For our purposes, it suffices to consider *simple* schedulers, in which for each state *s*, the scheduler fixes one of the enabled actions at *s* and selects the same action every time when the system resides in *s*. Formally, a simple scheduler is a function $\sigma : S \to Act$ such that $\sigma(s)$ is one of the actions enabled at state *s*. In the current setting, instead of considering the whole set of schedulers, we work only with a (finite) subset of simple schedulers Σ specified by the user. A *path* in $\mathcal{M}$ under σ is an infinite sequence of states $\rho = s_0 s_1 \cdots$ such that, for all $i \geq 0$, $\mathcal{P}_a[s_i, s_{i+1}] > 0$ for $a = \sigma(s_i)$. Let $Path_{\mathcal{M},\sigma}$ be the set of paths in $\mathcal{M}$ under σ. Let $Path_{\mathcal{M},\sigma}(s)$ be the subset of paths that start from s. Let $Pr_{\mathcal{M},\sigma}$ be the standard *probability distribution* over $Path_{\mathcal{M},\sigma}$ as defined in the literature [10, Ch. 10].

The *expected cumulative cost*, or simply *cumulative cost*, of reaching a set $G \subseteq S$ of *goal* states (called G-states hereafter) in $\mathcal{M}$ under σ, denoted $C_{\mathcal{M},\sigma}(G)$, is defined as follows: First, let $C_{\mathcal{M},\sigma}(s, G)$ be the expected value of random variable $X : Path_{\mathcal{M},\sigma}(s) \to \mathbf{R}_{\geq 0}$ such that (i) if $s \in G$ then $X(\rho) = 0$, (ii) if $\rho[i] \notin G$ for all $i \geq 0$ then $X(\rho) = \infty$, and (iii) otherwise $X(\rho) = \sum_{i=0}^{n-1} C(s_i)$ where $s_n \in G$ and $s_j \notin G$ for all $j < n$. Then, let $C_{\mathcal{M},\sigma}(G) = \sum_{s \in S} \alpha(s) \cdot C_{\mathcal{M},\sigma}(s, G)$.

By the above definitions, for those states which do *not* reach the goal states almost surely (viz. with probability less than 1), the cumulative cost is ∞. We remark that other definitions on the costs of paths not reaching the goal states do exist and can be found in [41]. However, they are more involved and are not needed in the current setting. In order to compute the cumulative cost, we first have to identify the set of states $S_{=1}$ from which the probability to reach the goal states in G is 1. This can be done by a standard graph analysis [10, Ch. 10]. Next, we solve the following system of linear equations with variables $(x_s)_{s \in S_{=1}}$:

$$
\begin{aligned}
x_s &= 0 & \text{if } s \in G \\
x_s &= C(s) + \sum_{t \in S_{=1}} \mathcal{P}_a[s,t] \cdot x_t & \text{if } s \notin G
\end{aligned}
\tag{9.1}
$$

where $a = \sigma(s)$. When the scheduler is fixed, the MDP is reduced to a discrete-time Markov chain (DTMC) and hence solving (9.1) is straightforward. One can employ standard Jacobi or Gauss-Seidel itertaion methods to compute the least fixpoint [179]. In detail, one starts from $\vec{x}^{(0)}$ where $x_s^{(0)} = 0$ for all $s \in S_{=1}$, and computes $x_s^{(n+1)} = C(s) + \sum_{t \in S_{=1}} \mathcal{P}_a[s,t] \cdot x_t^{(n)}$ if $s \notin G$ and 0; otherwise, until $\max_{s \in S} |x_s^{(n+1)} - x_s^{(n)}| < \varepsilon$ for some predetermined $\varepsilon > 0$. In practice, and especially in probabilistic verification, this is usually more efficient than the Gaussian elimination [70].

Interval-valued MDPs (IMDPs) are MDPs where some of the transition probabilities are specified as real intervals.

Definition 2 (IMDP) *An IMDP is a tuple* $\mathcal{M}^I = (S, Act, \mathcal{P}^+, \mathcal{P}^-, \alpha, C)$ *where*

- *S, Act, α, and C are defined the same as in Definition 1,*

- *$\mathcal{P}^+ = \{\mathcal{P}_a^+\}_{a \in Act}$, $\mathcal{P}^- = \{\mathcal{P}_a^-\}_{a \in Act}$ are two families of nonnegative matrices indexed by $a \in Act$, giving the* upper *and* lower *bounds of transition probabilities, respectively. Further, for each $a \in Act$, $\mathcal{P}_a^+$ and $\mathcal{P}_a^-$ have the same corresponding 0- and 1-entries.*

With $\mathcal{M}^I = (S, Act, \mathcal{P}^+, \mathcal{P}^-, \alpha, C)$ we associate a set of MDPs $[\![\mathcal{M}^I]\!]$ such that $\mathcal{M} = (S, Act, \mathcal{P}, \alpha, C) \in [\![\mathcal{M}^I]\!]$ *if and only if* for each $a \in Act$, $\mathcal{P}_a^- \leq \mathcal{P}_a \leq \mathcal{P}_a^+$ where $\leq$ is interpreted entry-wise. We call an $\mathcal{M} \in [\![\mathcal{M}^I]\!]$ an *instance* of $\mathcal{M}^I$.

Given an IMDP $\mathcal{M}^I$ and a simple scheduler σ, since the possible cumulative cost of reaching G-states is in the form of an interval, we are interested in the *bounds* of such an interval. The *minimum* cumulative cost of reaching G-states in $\mathcal{M}^I$ under σ is

$$C_{\mathcal{M}^I, \sigma}^{\min}(G) = \inf_{\mathcal{M} \in [\![\mathcal{M}^I]\!]} C_{\mathcal{M}, \sigma}(G).$$

Because the *maximum* cumulative cost $C_{\mathcal{M}^I, \sigma}^{\max}(G)$ is symmetrical to the minimum case, in the remainder of this section, we mainly deal with the latter.

To this end, as before we first identify states that reach the goal states G almost surely (under σ) and are denoted by $S_{=1}$. Owing to the assumption made on IMDPs in Definition 2, this can be done by graph-analysis as on MDPs $\mathcal{M}^I$. For those states not in $S_{=1}$, the minimal cost is ∞ according to our convention. One can consider the following Bellman equation over the variables $(x_s)_{s \in S_{=1}}$:

$$
\begin{aligned}
x_s &= 0 && \text{if } s \in G \\
x_s &= \min_{\mathcal{P}_a^- \leq \mathcal{P}_a \leq \mathcal{P}_a^+} \Big\{ C(s) + \sum_{t \in S_{=1}} \mathcal{P}_a[s,t] \cdot x_t \Big\} && \text{if } s \notin G
\end{aligned}
\tag{9.2}
$$

where $a = \sigma(s)$. Note that $\mathcal{P}_a$ is required to be a transition probability matrix. Let $\vec{x} = (x_s)_{s \in S_{=1}}$ be the *least* fixpoint of (9.2). We easily obtain:

Proposition 1 $C_{\mathcal{M}^I, \sigma}^{\min}(G) = \sum_{s \in S} \alpha(s) x_s.$

To solve (9.2), there are essentially two approaches. The first one is to reduce it to linear programming (LP). However, despite theoretically elegant, this is not practical for real-life cases. Instead, we apply the second approach, *i.e.*, the value-iteration method. For each iteration, the crucial part is to compute

$$\min_{\mathcal{P}_a^- \leq \mathcal{P}_a \leq \mathcal{P}_a^+} \Big\{ C(s) + \sum_{t \in S_{=1}} \mathcal{P}_a[s,t] \cdot x_t \Big\}$$

for a given $\vec{x}$. This problem can be reduced to a standard linear program. Indeed, for each s, introduce variables $(y_t)_{t \in S}$ and consider the problem:

$$\text{minimize} \quad C(s) + \sum_{t \in S_{=1}} y_t x_t$$

$$\text{subject to} \quad \sum_{t \in S_{=1}} y_t = 1 \text{ and } \mathcal{P}_a^-[s,t'] \leq y_{t'} \leq \mathcal{P}_a^+[s,t'] \text{ for all } t' \in S_{=1}.$$

This can be solved efficiently via off-shelf LP solvers (note that here x_t's and a are given). Hence, each iteration takes polynomial time. We also remark that the LP here admits a very simple structure and only contains at most $|S|$ variables (and usually much less for practical examples), while the direct approach (based on LP as well) requires at least $|S|^2 + |S|$ variables and is considerably more involved. Although it might take exponentially many iterations to reach the least fixpoint, in practice, one usually sets a stopping criteria such as $\max_{s \in S} |x_s^{(n+1)} - x_s^{(n)}| < \varepsilon$ for a fixed error bound $\varepsilon > 0$.

Let $C_{\mathcal{M}^I,\sigma}^{\text{dif}}(G) = C_{\mathcal{M}^I,\sigma}^{\max}(G) - C_{\mathcal{M}^I,\sigma}^{\min}(G)$. Because $\mathcal{M}$, $\mathcal{M}^I$ and G are clear in context, to simplify notations, we make the following abbreviations:

FULLY-SPELLED	$C_{\mathcal{M},\sigma}(G)$	$C_{\mathcal{M}^I,\sigma}^{\min}(G)$	$C_{\mathcal{M}^I,\sigma}^{\max}(G)$	$C_{\mathcal{M}^I,\sigma}^{\text{dif}}(G)$
ABBREVIATED	C_σ	$C_\sigma^{\min}$	$C_\sigma^{\max}$	C_σ^{dif}

9.2.2 An overview of IDMS

IDMS is an iterative process that contains one pre-stage and five runtime steps. The pre-stage builds up a parametric MDP with transition probability parameters in the design time. At runtime,

1. Step 1: collect data samples;

2. Step 2: infer point and interval estimates based on the samples;

3. Step 3: build up a (concrete) MDP and an IMDP, by instantiating the parameters with the point and interval estimates;

4. Step 4: attempt to compute a confidently optimal scheduler; and

5. Then the process either moves to Step 5 where a decision is made or goes back to Step 1.

The process terminates when either a confidently optimal scheduler is returned or the maximal time of iteration (namely the maximal number of steps within the iteration) is reached. Note as the decision-making may need to be repeated periodically at runtime, Step 5 may be followed by Step 1.

Example 1 *The following example is given in [169] as a running example, where a parametric MDP $\mathcal{M}_{eg}(\vec{\theta})$ is described in Figure 9.1. The state space of $\mathcal{M}_{eg}(\vec{\theta})$ is $\{s_0, \ldots, s_7, s_G\}$ with s_0 being the only initial state (i.e., the initial distribution assigns 1 to s_0 and 0 to other states) and s_G being the only goal state. The dashed arrows are probabilistic transitions, labeled by parameters $\vec{\theta} = (\theta_1, \ldots, \theta_5)$. The solid arrows are non-probabilistic transitions (or, equivalently, transitions with the fixed probability 1). The wavy arrows represent non-deterministic transitions, with a and b being two actions. For $\mathcal{M}_{eg}(\vec{\theta})$, the two actions induce two schedulers, denoted σ_a and σ_b, respectively. States of $\mathcal{M}_{eg}(\vec{\theta})$ are associated with costs ranging from 0 to 2.*

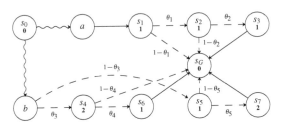

Figure 9.1: A parametric MDP example $\mathcal{M}_{eg}(\vec{\theta})$

It is worth mentioning that IDMS does not presume a particular method for collecting runtime data. However, when such data is collected, IDMS employs a simple data structure, namely a set of non-negative integer matrices, to store them. These matrices are related to schedulers of the parametric MDP. The integer in each entry represents the number of times that the corresponding transition is recorded in the sampling time frame. For example, the two integer matrices related to σ_a and σ_b of $\mathcal{M}_{eg}(\vec{\theta})$ are as follows:

$$
M_a: \begin{array}{c} s_1 \\ s_2 \end{array}
\begin{array}{cccc} s_1 & s_2 & s_3 & s_G \end{array}
\begin{bmatrix} 0 & N_{1,2} & 0 & N_{1,G} \\ 0 & 0 & N_{2,3} & N_{2,G} \end{bmatrix}
\qquad
M_b: \begin{array}{c} s_0 \\ s_4 \\ s_5 \end{array}
\begin{array}{ccccc} s_4 & s_5 & s_6 & s_7 & s_G \end{array}
\begin{bmatrix} N_{0,4} & N_{0,5} & 0 & 0 & \\ 0 & 0 & N_{4,5} & 0 & N_{4,G} \\ 0 & 0 & 0 & N_{5,7} & N_{5,G} \end{bmatrix}
$$

where $N_{\sharp,\natural} > 0$, with $\sharp$ and $\natural$ denoting (some) elements in $\{0,\dots,7,G\}$, are integer variables. $N_{\sharp,\natural}$ is increased by 1 (*i.e.*, $N_{\sharp,\natural} \leftarrow N_{\sharp,\natural} + 1$) if a transition from $s_\sharp$ to $s_\natural$ is newly observed. Note that zero entries in M_a and M_b remain unchanged for all time, because according to the structural specification of $\mathcal{M}_{eg}(\vec{\theta})$, the correspondent transitions are impossible to occur.

The data structure is used to estimate parameters in the parametric MDP. IDMS adopts two forms of estimation, namely *point estimation* and *interval estimation*, which are illustrated by M_a. Note that M_a is used to estimate parameters θ_1 and θ_2. For point estimation, θ_1 is estimated as the numerical value $N_{1,2}/(N_{1,2} + N_{1,G})$, and θ_2 is estimated as $N_{2,3}/(N_{2,3} + N_{2,G})$. For interval estimation, IDMS assumes that θ_1 (resp., θ_2) is the mean of a Bernoulli distribution and $(N_{1,2}, N_{1,G})$ (resp., $(N_{2,3}, N_{2,G})$) forms a random sample of the distribution. In other words, $(N_{1,2}, N_{1,G})$ denote a random sample containing $N_{1,2}$ copies of 1 and $N_{1,G}$ copies of 0, and $(N_{2,3}, N_{2,G})$ has a similar meaning. Therefore, one can employ the standard statistical inference method to derive a confidence interval for θ_1 and one for θ_2. By the law of large numbers, if $N_{1,2} + N_{1,G}$ (resp., $N_{2,3} + N_{2,G}$) increases, then the width of the resulted confidence interval for θ_1 (resp., θ_2) likely decreases (when the confidence level is fixed).

9.2.3 Confident optimality

By instantiating the transition probability parameters in the parametric MDP with the corresponding point estimates and interval estimates, one obtains a concrete MDP $\mathcal{M}$ and an IMDP $\mathcal{M}^I$. Note that if $[p,q] \subset [0,1]$ instantiates a parameter θ then, equivalently, $[1-q, 1-p]$ instantiates $1-\theta$. Clearly, $\mathcal{M}$ and $\mathcal{M}^I$ share the same state space S, initial distribution α, and cost function C. Moreover, $\mathcal{M}$ is an instance of $\mathcal{M}^I$, namely, $\mathcal{M} \in [\![\mathcal{M}^I]\!]$. From now on, for given $\mathcal{M}$ and $\mathcal{M}^I$, we always assume $\mathcal{M} \in [\![\mathcal{M}^I]\!]$. A key decision-making criterion in IDMS is formalized as follows:

Definition 3 (Confident optimality) *Given* $\mathcal{M}$, $\mathcal{M}^I$, $G \subseteq S$ *of goal states and a finite nonempty subset* Σ *of schedulers,* $\sigma^* \in \Sigma$ *is confidently optimal if, for all* $\sigma \in \Sigma \backslash \sigma^*$, *the following two conditions hold:*

$$C_{\sigma^*} \leq C_{\sigma}, \text{ and}$$
$$C_{\sigma^*}^{\max} \leq C_{\sigma}^{\min} + \gamma \cdot C_{\sigma^*}^{\text{dif}} \text{ where } \gamma \geq 0. \tag{9.3}$$

In words, a scheduler σ^* in the given scheduler subset Σ of $\mathcal{M}$ (or, equivalently, $\mathcal{M}^I$) is confidently optimal if for *all other* schedulers σ in Σ (*i.e.*, $\sigma \neq \sigma^*$):

■ the cumulative cost (of reaching G-states) in $\mathcal{M}$ under σ^* is not larger than the cumulative cost in $\mathcal{M}$ under σ; and

■ the $(1/\gamma)$-portion of the difference between the maximum cumulative cost in $\mathcal{M}^I$ under σ^* and the minimum cumulative cost in $\mathcal{M}^I$ under σ is not larger than the maximum-minimum difference of cumulative cost in $\mathcal{M}^I$ under σ^*.

A correct illustrative example is presented in the latter text. It is noteworthy that, different from a standard MDP problem, a subset of schedulers is explicitly given in our definition.

The parameter γ, which is specified by the user, has the function of adjusting the criterion of confident optimality. A confidently optimal scheduler may not exist for the given MDP and IMDP; in some rare case, there may be more than one confidently optimal schedulers. Note that if a sufficiently large value for γ is selected, then the second condition in Equation (9.3) is guaranteed to be true. If so, the definition is degenerated to the standard definition of optimal cumulative costs for MDPs with point-valued transition probabilities.

Given $\mathcal{M}, \mathcal{M}^I, G, \Sigma, \gamma$, the following procedure decides whether a confidently optimal scheduler σ^* exists and returns σ^* if it exists:

1. Compute C_σ for all $\sigma \in \Sigma$, and compute $\Sigma_1 \subseteq \Sigma$ such that $C_{\sigma_1} = \min_{\sigma \in \Sigma} C_\sigma$, if and only if, $\sigma_1 \in \Sigma_1$;

2. Compute $C_{\sigma_1}^{\max}$ for all $\sigma_1 \in \Sigma_1$, and compute $C_\sigma^{\min}$ for all $\sigma \in \Sigma$; and

3. If there is $\sigma^* \in \Sigma_1$ such that $C_{\sigma^*}^{\max} \leq C_{\sigma}^{\min} + \gamma \cdot C_{\sigma^*}^{\dif}$ where $\sigma \neq \sigma^*$, then return σ^*; otherwise, return "no confidently optimal scheduler."

The procedure relies on the core method of value-iteration presented in Section 9.2.1. The computational complexity is dependent on the core value-iteration method and the size of Σ. Note that although the number of *all* schedulers in an MDP increases exponentially as the size of the MDP increases, in our case, a *specific* subset of schedulers Σ is predefined by the model builder. If we suppose the value-iteration takes constant time (*e.g.*, the model is fixed), then the time complexity of the procedure is linear in the size of Σ.

We present an example to explain how IDMS is affected by γ and the sample size. Suppose after instantiating $\vec{\theta}$ of $\mathcal{M}_{eg}(\vec{\theta})$ with point estimates and interval estimates, the cumulative cost intervals for schedulers σ_a and σ_b are $[l_1, u_1]$ and $[l_2, u_2]$, respectively. The positions of $l_1, u_1, l_2,$ and u_2 are illustrated on the left side of the following drawing (where $0 \leq p < q$).

If $u_1 \leq l_2 + \gamma(u_1 - l_1)$, the above procedure returns σ_a. But if $u_1 > l_2 + \gamma(u_1 - l_1)$, neither σ_a nor σ_b is confidently optimal, and the procedure returns "no confidently optimal scheduler." If one lowers the value γ and/or increases the sample size, the computed cost intervals usually shrink, as depicted on the right side of the above drawing. Then there is a higher probability that a confidently optimal scheduler (namely σ_a) is returned from the procedure and the iteration of IDMS terminates.

9.2.4 Metrics and tradeoff

One main advantage of IDMS is the flexibility that enables a tradeoff between the three important metrics for practical, especially runtime, decision-making. The three metrics are accuracy of the decision, data usage for making the decision, and computational overhead on the runtime system. Because random sampling is involved in IDMS, under a specific scheduler of an MDP and an IMDP, the cumulative cost and the minimum/maximum cumulative costs (of reaching the goal states) are uncertain. Therefore, a confidently optimal scheduler may be decided at each iterative step with a certain probability. Further, a confidently optimal scheduler may not be the truly optimal one, which is defined based on the unknown real values of the transition probability parameters in the abstract MDP. In view of this, we define the three metrics as follows:

■ accuracy is the probability that a confidently optimal scheduler is optimal;

■ data usage is the average size of sampled data used in the iteration; and

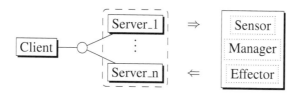

Figure 9.2: Software architecture of Z.com

■ computational overhead is measured by the average iteration time (namely, the average number of iterative steps).

Ideally, one wants to maximize the first one while minimize the latter two. However, according to laws of statistics, this is impossible. To obtain high accuracy in a statistical process (including IDMS), a large-sized sample has to be used; although it is possible to set a high accuracy threshold and then try to infer the result using a sample whose size is as small as possible, this usually leads to a costly iterative process. Therefore, a practical solution is to achieve a suitable tradeoff between the three metrics. In IDMS, to realize this tradeoff, one can adjust the constant γ and the sample size within the iteration.

9.3 Application to Self-Adaptive Systems

In this section, we describe an application of IDMS to self-adaptive systems, in particular, the Rainbow framework.

9.3.1 Rainbow framework

We illustrate Rainbow with the example Z.com [45] which is a fictional news website providing multi-media and textual news service while keeping the cost of maintaining its server pool within its operational budget. Z.com has a client-server architecture with three additional adaptation-relevant components, as shown in Figure 9.2: the Sensor collects runtime data; the Manager controls the adaptation, such as switching the news content mode from multi-media to text and *vice versa*; and the Effector executes the adaptation to affect the system.

In Rainbow, the adaptation is specified as *strategies* in its customized language Stitch [42, 44]. A strategy is a tree structure consisting of *tactics*, which in turn contain operations. Figure 9.3 specifies two strategies a and b, guarded by a common condition cond where SNo and MaxSNo refer to the current server number and the maximal server number, respectively. If strategy a is selected, operation enlistSever[1] in tactic s1 is first executed. Next, if the variable hiLatency is true then enlistSever[1] in tactic s2 is executed; otherwise, strategy a

---

For simplicity, the specification does not strictly follow the syntax of Stitch.

```
define cond:=hiLatency&!TextMode&(SNo<=MaxSNo−2);
strategy a[cond]{
  tactic s1:enlistServer[1]{
    tactic s2:hiLatency−>enlistServer[1]{
      tactic s3:hiLatency−>switchToTextMode;}}}
strategy b[cond]{
  tactic s4:hiLoad−>enlistServer[2]{
    tactic s6:hiLatency−>switchToTextMode;}
  tactic s5:!hiLoad−>switchToTextMode{
    tactic s7:hiLatency−>enlistServer[2];}}
... % other strategy specification
```

Figure 9.3: Strategy specification for Z.com in Stitch

Table 9.1: Costs of operations in strategies a and b

utility dimension	operation						
	op(s1)	op(s2)	op(s3)	op(s4)	op(s5)	op(s6)	op(s7)
content	0	0	1	0	1	1	0
budget	1	1	0	2	0	0	2

terminates. Last, if hiLatency persists to be true then switchToTextMode in tactic s3 is executed; otherwise, strategy a terminates. Strategy b is specified in a similar style.

To evaluate strategies, Rainbow uses *utilities* to describe the costs and benefits of operations. The quantities of utilities are provided by human experts or stakeholders. Table 9.1 describes two utilities called *content* and *budget* and the costs of the operations in terms of the two. Note that because there is only one operation in each tactic of the adaptation specification in Figure 9.3, we use tactic names to label operations—the correspondent operation to a tactic s is denoted op(s). For example, if switchToTextMode is executed, then the content cost is, say, 1; if enlistServer[i] with $i \in \{1,2\}$ is executed, then the budget cost is, say, i. Then, the overall cost of an operation is the weighted sum of utilities. For simplicity, we let the weights of all utilities be equal to 1.

Rainbow characterizes uncertainty in the detection of guarding conditions (such as hiLantency in tactic s2) as probabilities called *likelihoods*. The likelihoods in strategies a and b are specified in Table 9.2. Note that because there is one likelihood parameter in each tactic (except s1) in Figure 9.3, like for operations, we also use tactic names to label likelihoods—the correspondent operation to a tactic s is denoted lk(s). We explain how these likelihoods are elicited in Rainbow later; for now, they are viewed as undetermined parameters.

The correspondence between the adaptation specification of Z.com and an MPD model is straightforward. Namely, operations are represented by actions and strategies are represented by schedulers. Indeed, the Stitch specification under consideration can be translated into $\mathcal{M}_{eg}(\vec{\theta})$. Therefore, the adaptation prob-

Table 9.2: Likelihood parameters in strategies a and b

likelihood	interpretation as a conditional probability
lk(s2)	Pr(hiLatency=true | SNo=MaxSNo - 1 & textMode=true)
lk(s3)	Pr(hiLatency=true | SNo=MaxSNo & textMode=true)
lk(s4)	Pr(hiLoad=true | hiLatency=true & SNo=MaxSNo - 2 & textMode=true)
lk(s4)	Pr(hiLoad=false | hiLatency=true & SNo=MaxSNo - 2 & textMode=true)
lk(s6)	Pr(hiLatency=true | hiLoad=true & SNo=MaxSNo & textMode=true)
lk(s7)	Pr(hiLatency=true | hiLoad=true & SNo=MaxSNo - 2 & textMode=false)

lem in Rainbow is an instance of the problem of selecting a strategy that minimizes the cumulative cost (of reaching the goal states in the MDP).

9.3.2 Embedding IDMS into rainbow

Rainbow supports at least two methods to elicit likelihoods. First, like utilities and their weights, concrete values of likelihoods can be explicitly given by human experts or stakeholders [42]. Second, sampling methods for estimating likelihoods are also implemented in Rainbow [38, 43]. For example, the Manager can check the values of Boolean variables hiLatency and hiLoad as the system operates and record the result. Then, with respect to the condition probabilities described in Table 9.2, one easily obtains a sample for each parameter θ_i. Therefore, we can embed IDMS into Rainbow economically, just by enhancing the reasoning mechanism of strategy selection in the Manager with IDMS, but with little change made to the Sensor and the Effector.

Rainbow exploits point estimates for likelihoods, as its decision model is a standard MDP. Because the runtime data set cannot be arbitrarily large, point estimates may be error-prone. Poor strategy selection often causes some extra cost and reduced benefit. Even worse, the extra cost and reduced benefit may accumulate if the non-optimal strategy is selected repeatedly. In light of this, the interval estimation method in IDMS is used to complement to the point estimation method in Rainbow and leads to more stable decision-making outputs. By applying IDMS to Rainbow, another (and perhaps more important) benefit is the possibility of making a tradeoff between accuracy, data usage, and computational overhead, thus, improving the adaptation function of Rainbow.

9.3.3 Experiments

Recall that IDMS assumes that likelihood parameters in Z.com are means of Bernoulli distributions. We use Matlab to simulate the generation and collection of runtime data. To this end, we need to fix the expected values of the Bernoulli random variables, namely the true values of $\vec{\theta}$ of $\mathcal{M}_{eg}(\vec{\theta})$. We let $\theta_1 = \frac{2}{3}$, $\theta_2 = \frac{4}{7}$, $\theta_3 = \frac{1}{3}$, $\theta_4 = \frac{4}{9}$, and $\theta_5 = \frac{4}{9}$. As the true values of $\vec{\theta}$ are given, we also know

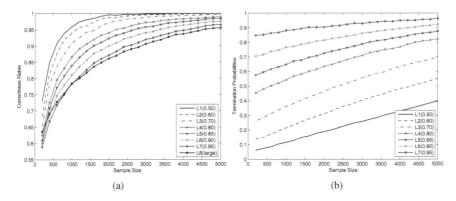

Figure 9.4: (a) Correctness rates and (b) termination probabilities with different sample sizes and γ values

which scheduler is optimal. Indeed, by computation, the overall cost of strategy a is 2.0476 and that of strategy b is 2.0741. Thus, strategy a is optimal. It is noteworthy that the difference between the above two overall costs may seem small, but it is non-negligible because they are proportional to the weights of utility dimensions, which may be large in some case, and also because the extra cost may accumulate if the adaptation is triggered repeatedly.

To evaluate the flexibility of IDMS for making the intended tradeoff, we implement the computing procedure presented in Section 9.2.3 in Matlab. Given a sample of specific size for estimating each parameter θ_i of $\vec{\theta}$, and given a specific value of γ, IDMS terminates with a certain probability called *termination probability* in the experiment. Based on the termination probability, we can immediately calculate the data usage and the computational overhead. Upon termination, with a certain probability, the selected scheduler is strategy a. This probability, called *correctness rate* in the experiment, is equal to the metric of accuracy. Since we can simulate IDMS, (applied to $\mathcal{M}_{eg}(\vec{\theta})$), we can estimate the correctness rate and termination probability using the standard Monte Carlo estimation. In this experiment, we estimate the two for *different* sample sizes and values of γ. Note that the confidence level of interval estimation is fixed in IDMS and we set it to be 95% in the experiment.

The experimental data, summarized in Figure 9.4, is generated from samples of n-size with n ranging from 200 to 5,000 in increments of 200, and with a selection of values for γ as specified in the legends of the figures (where "large" refers to a sufficiently large value of γ such that the computing procedure is degenerated to a point estimation). For each n, the number of generalized samples is 10,000, based on which we calculate the correctness rate and termination probability.

Figure 9.4 demonstrates the dependence of the correctness rate and termination probability on γ and the sample size. Figure 9.4(a) shows that as γ decreases

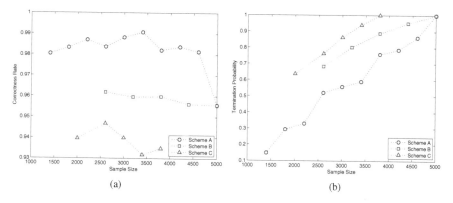

(a)   (b)

Figure 9.5: Three iteration schemes in items of (a) correctness rates and (b) termination probabilities

or as the sample size increases, the correctness rate increases. In particular, except for samples of small size (less than 1,000), IDMS provides a higher correctness rate than the point estimation method. Figure 9.4(b) shows that as γ increases or as the sample size increases, the termination probability increases. Note that if a sufficiently large value for γ is selected, the termination probability is 1 for samples of all selected sizes (and, thus, L8 is not depicted in Figure 9.4(b).)

An important implication of Figure 9.4 is that, by adjusting the value of γ and the sample size in different ways, one is able to achieve different tradeoffs between accuracy, data usage, and computational overhead. To illustrate this flexibility, Table 9.3 describes three cases where the three metrics have different priorities. Based on Figure 9.4, by selecting different pairs of γ and sample size, we obtain three examples of iteration schemes depicted in Figure 9.5. Each marker in Figure 9.5 refers to an iterative step with a specific value of γ and a specific sample size. For example, setting $\gamma = 0.5$ and the sample size as 1,400, according to Figure 9.4, we obtain the leftmost marker of Scheme A in Figure 9.5. The other markers in Figure 9.5 are identified in the same way. All three schemes terminate with probability 1 before or when the sample size reaches 5,000. It is easy to observe that the schemes reflect the metric priorities in the corresponding

Table 9.3: Priorities of metrics in three different cases

metric	priority		
	A	B	C
accuracy	high	medium	low
data usage	low	medium	high
computational overhead	low	high	high

cases in Table 9.3. For example, Scheme A has a high correctness rate compared with the other two schemes, because the priority of accuracy is high in Case A; it has a low average termination probability and a high number of markers, because the priorities of both data usage and computational overhead are low in Case A.

9.4 Further Reading

Several high-level frameworks and approaches based on probabilistic model checking have been proposed for self-adaptive systems recently, but with emphasis on different aspects of the adaptation, such as QoS management and optimization [34], adaptation decisions [80], verification with information of confidence intervals [32], runtime verification efficiency and sensitivity analysis [66], and proactive verification and adaptation latency [133]. None of those works addressed the problem of making a practical tradeoff similar to the one supported by IDMS. Rainbow [42] supports the computation of cumulative costs and/or rewards when the likelihood parameters in the adaptation strategies are explicitly specified. Subsequent work [35, 36] employs a combination of a simulation method and probabilistic model checking to evaluate properties such as resilience and adaptation latency. As mentioned, our IDMS can be economically embedded into Rainbow and extend the adaptation function of the latter. Other aspects of the self-adaptation uncertainty are also modeled probabilistically, such as model drift [59], reliability prediction [46], and resource availability [144].

It is worth mentioning some other existing approaches to the design of self-adaptive systems, which rely on mathematical methods related to probability theory and statistics. Esfahani *et al.* [63, 64] presented a general definition of adaptation optimality using fuzzy mathematics, which accounts for not only the current utility but also the optimal consequence of future operations. But IDMS estimates the probability parameters based on runtime data. Epifani *et al.* [62] presented the KAMI framework to deal with the inaccuracy of parameters related to the non-functional aspect of the system (such as reliability and performance), and Bencomo *et al.* [21] presented a Bayesian network for modeling self-adaptive systems. These two approaches rely on the Bayesian (point) estimation method while IDMS exploits both point and interval estimates from the frequentist statistics theory. Moreover, they considered the verification of requirements, while we target the computation of cumulative costs, but their approach parameters are elicited by point estimates. Letier *et al.* [117] presented a methodology for making architectural design decisions (and, thus, is broader than the self-adaptation domain), in which Monte Carlo simulations are used to elicit probability distributions of parameters in the model. Their method targets static decision-making, and does not use dynamic models for reasoning the statistical estimates. Finally, Filieri *et al.* [68] constructed approximate dynamic models of a self-adaptive system and for synthesizing, from those models, a suitable controller that guarantees prescribed multiple non-functional system requirements. The method they used is from control theory, which is quite different from the theory of MDPs.

9.5 Summary

Software in the open environment is often governed by and thus adapts to phenomena that occur at runtime. Unlike traditional decision problems, where a decision-making model is determined for reasoning, the adaptation logic of such software is concerned with empirical data and is subject to practical constraints. In this chapter, we give an introduction to IDMS, an iterative framework that supports a tradeoff among three important metrics in practical runtime decision-making problems: accuracy, data usage, and computational overhead. IDMS can infer both point and interval estimates for the undetermined transition probabilities in an MDP based on sampled data, and iteratively computes a confidently optimal scheduler from a given finite subset of schedulers. The most important feature of IDMS is the flexibility for adjusting the criterion of confident optimality and the sample size within the iteration, leading to a tradeoff between accuracy, data usage, and computational overhead. We apply IDMS to an existing self-adaptation framework, Rainbow, and conduct a case study using a Rainbow system to demonstrate the flexibility of IDMS. As an example to demonstrate its applicability to self-adaptive systems, IDMS is instantiated on the Rainbow framework with a simulation-based evaluation.

References

[1] Introduction to cloud computing architecture. *White Paper, Sun Microsystems, Inc.*, June, 2009.

[2] Gregory Abowd, Anind Dey, Peter Brown, Nigel Davies, Mark Smith, and Pete Steggles. Towards a better understanding of context and context-awareness. In *International Symposium on Handheld and Ubiquitous Computing*, pages 304–307. Springer, 1999.

[3] Pekka Abrahamsson, Outi Salo, Jussi Ronkainen, and Juhani Warsta. Agile software development methods. *Review and Analysis VTT Publication*, 478, 2002.

[4] Sarel Aiber, Dagan Gilat, Ariel Landau, Natalia Razinkov, Aviad Sela, and Segev Wasserkrug. Autonomic self-optimization according to business objectives. In *Autonomic Computing, 2004. Proceedings. International Conference on*, pages 206–213, 2004.

[5] Rajeev Alur and David Dill. A theory of timed automata. *Theoretical computer science*, 126(2):183–235, 1994.

[6] David Anderson, Jeff Cobb, Eric Korpela, Matt Lebofsky, and Dan Werthimer. SETI@ home: an experiment in public-resource computing. *Communications of the ACM*, 45(11):56–61, 2002.

[7] Grigoris Antoniou and Frank Van Harmelen. Web ontology language: Owl. In *Handbook on Ontologies*, pages 91–110. Springer, 2009.

[8] Assaf Arkin, Sid Askary, Scott Fordin, Wolfgang Jekeli, Kohsuke Kawaguchi, David Orchard, Stefano Pogliani, Karsten Riemer, Susan Struble, Pal Takacsi-Nagy, et al. Web service choreography interface (WSCI) 1.0. *Standards Proposal by BEA Systems, Intalio, SAP, and Sun Microsystems*, 2002.

[9] Michael Armbrust, Armando Fox, Rean Griffith, Anthony Joseph, Randy Katz, Andrew Konwinski, Gunho Lee, David Patterson, Ariel Rabkin, Ion Stoica, et al. Above the clouds: A Berkeley view of cloud computing. *EECS Department, University of California, Berkeley, Tech. Rep. UCB/EECS-2009-28*, 2009.

[10] Christel Baier and Joost-Pieter Katoen. *Principles of Model Checking.* The MIT Press, 2008.

[11] Mark Baker, Rajkumar Buyya, and Domenico Laforenza. The Grid: International efforts in global computing. In *Proceedings of the International Conference on Advances in Infrastructure for Electronic Business, Science, and Education on the Internet*, 2000.

[12] Henri Bal, Raoul Bhoedjang, Rutger Hofman, Ceriel Jacobs, Thilo Kielmann, Jason Maassen, Rob Van Nieuwpoort, John Romein, Luc Renambot, Tim Rühl, et al. The distributed ASCI supercomputer project. *ACM SIGOPS Operating Systems Review*, 34(4):76–96, 2000.

[13] Matthias Baldauf, Schahram Dustdar, and Florian Rosenberg. A survey on context-aware systems. *International Journal of Ad Hoc and Ubiquitous Computing*, 2(4):263–277, 2007.

[14] H. Balen. *Distributed Object Architectures with CORBA.* Cambridge University Press, 2000.

[15] Luciano Baresi, Elisabetta Di Nitto, and Carlo Ghezzi. Toward open-world software: Issue and challenges. *Computer*, 39(10):36–43, 2006.

[16] Luciano Baresi, Carlo Ghezzi, Xiaoxing Ma, and Valerio Panzica La Manna. Efficient dynamic updates of distributed components through version consistency. *IEEE Transactions on Software Engineering*, 2016.

[17] Luciano Baresi, Carlo Ghezzi, and Luca Mottola. Loupe: Verifying publish-subscribe architectures with a magnifying lens. *Software Engineering, IEEE Transactions on*, 37(2):228–246, 2011.

[18] Luciano Baresi and Reiko Heckel. Tutorial introduction to graph transformation: A software engineering perspective. In *Proc. of the First International Conference on Graph Transformation, ICGT 2002*, pages 402–429. Springer, 2002.

[19] Len Bass, Ingo Weber, and Liming Zhu. *DevOps: A Software Architect's Perspective.* Addison-Wesley Professional, 2015.

[20] Gerd Behrmann, Alexandre David, and Kim G Larsen. A tutorial on uppaal. In *Formal Methods for the Design of Real-Time Systems*, pages 200–236. Springer, 2004.

[21] Nelly Bencomo, Amel Belaggoun, and Valerie Issarny. Dynamic decision networks for decision-making in self-adaptive systems: A case study. In *Proceedings of the 8th International Symposium on Software Engineering for Adaptive and Self-Managing Systems*, SEAMS'13, pages 113–122, Piscataway, NJ, USA, 2013. IEEE Press.

[22] Gérard Berry and Gérard Boudol. The chemical abstract machine. In *Proceedings of the 17th ACM SIGPLAN-SIGACT Symposium on Principles of Programming Languages*, pages 81–94. ACM New York, NY, USA, 1989.

[23] Christian Bettstetter and Christoph Renner. A comparison of service discovery protocols and implementation of the service location protocol. In *Proc. of the 6th EUNICE Open European Summer School: Innovative Internet Applications*, pages 13–15, 2000.

[24] Joseph Bigus, Don Schlosnagle, Jeff Pilgrim, Nathaniel Mills, and Yixin Diao. Able: A toolkit for building multiagent autonomic systems. *IBM Systems Journal*, 41(3):350–371, 2002.

[25] Antonis Bikakis, Theodore Patkos, Grigoris Antoniou, and Dimitris Plexousakis. A survey of semantics-based approaches for context reasoning in ambient intelligence. *Constructing Ambient Intelligence: 2007 Workshops, Germany, 2007*, page 14.

[26] Andrew Birrell and Bruce Jay Nelson. Implementing remote procedure calls. *ACM Transactions on Computer Systems (TOCS)*, 2(1):39–59, 1984.

[27] IEEE Standards Board. IEEE standard glossary of software engineering terminology. *IEEE Std*, 610.12-1990(121990), Approved 1990, Reaffirmed 2002.

[28] Jeremy Bradbury, James Cordy, Juergen Dingel, and Michel Wermelinger. A survey of self-management in dynamic software architecture specifications. In *Proceedings of the 1st ACM SIGSOFT Workshop on Self-Managed Systems*, pages 28–33. ACM, New York, NY, USA, 2004.

[29] Yérom-David Bromberg and Valérie Issarny. INDISS: Interoperable discovery system for networked services. *Lecture Notes in Computer Science*, 3790:164, 2005.

[30] F.P. Brooks. No silver bullet: essence and accidents of software engineering. *IEEE Computer*, 20(4):10–19, 1987.

[31] Peter Brown, John Bovey, and Xian Chen. Context-aware applications: From the laboratory to the marketplace. *IEEE [see also IEEE Wireless Communications] Personal Communications*, 4(5):58–64, 1997.

[32] Radu Calinescu, Carlo Ghezzi, Kenneth Johnson, Mauro Pezzé, Yasmin Rafiq, and Giordano Tamburrelli. Formal verification with confidence intervals: A new approach to establishing the quality-of-service properties of software systems. *Reliability, IEEE Transactions on*, 2015.

[33] Radu Calinescu, Carlo Ghezzi, Marta Kwiatkowska, and Raffaela Mirandola. Self-adaptive software needs quantitative verification at runtime. *Communications of the ACM*, 55(9):69–77, 2012.

[34] Radu Calinescu, Lars Grunske, Marta Kwiatkowska, Raffaela Mirandola, and Giordano Tamburrelli. Dynamic QoS management and optimization in service-based systems. *Software Engineering, IEEE Transactions on*, 37(3):387–409, 2011.

[35] Javier Cámara and Rogério de Lemos. Evaluation of resilience in self-adaptive systems using probabilistic model-checking. In *Software Engineering for Adaptive and Self-Managing Systems (SEAMS), 2012 ICSE Workshop on*, pages 53–62, June 2012.

[36] Javier Cámara, Gabriel A. Moreno, and David Garlan. Stochastic game analysis and latency awareness for proactive self-adaptation. In *Proceedings of the 9th International Symposium on Software Engineering for Adaptive and Self-Managing Systems*, SEAMS 2014, pages 155–164, ACM, New York, NY, USA, 2014.

[37] Luca Cavallaro, Elisabetta Di Nitto, and Matteo Pradella. An automatic approach to enable replacement of conversational services. In *Service-Oriented Computing*, pages 159–174. Springer, 2009.

[38] Orieta Celiku, David Garlan, and Bradley Schmerl. Augmenting architectural modeling to cope with uncertainty. In *Proceedings of the International Workshop on Living with Uncertainty (IWLU'07)*, Atlanta, Georgia, USA, 2007.

[39] Haibo Chen, Jie Yu, Chengqun Hang, Binyu Zang, and Pen-Chung Yew. Dynamic software updating using a relaxed consistency model. *Software Engineering, IEEE Transactions on*, 37(5):679–694, 2011.

[40] Harry Chen, Tim Finin, and Anupam Joshi. An ontology for context-aware pervasive computing environments. *The Knowledge Engineering Review*, 18(03):197–207, 2004.

[41] Taolue Chen, Vojtech Forejt, Marta Z. Kwiatkowska, David Parker, and Aistis Simaitis. Automatic verification of competitive stochastic systems. *Formal Methods in System Design*, 43(1):61–92, 2013.

[42] Shang-Wen Cheng. *Rainbow: Cost-Effective Software Architecture-based Self Adaptation*. PhD thesis, Carnegie Mellon University, 2008.

[43] Shang-Wen Cheng and D Garlan. Handling uncertainty in autonomic systems. In *Proceedings of the International Workshop on Living with Uncertainty (IWLU'07)*, Atlanta, Georgia, USA, 2007.

[44] Shang-Wen Cheng and David Garlan. Stitch: A language for architecture-based self-adaptation. *Journal of Systems and Software*, 85(12):2860–2875, 2012.

[45] Shang-Wen Cheng, David Garlan, and Bradley Schmerl. Architecture-based self-adaptation in the presence of multiple objectives. In *Proceedings of the 2006 International Workshop on Self-Adaptation and Self-Managing Systems*, pages 2–8. ACM, New York, NY, USA, 2006.

[46] Deshan Cooray, Sam Malek, Roshanak Roshandel, and David Kilgore. RESISTing reliability degradation through proactive reconfiguration. In *Proceedings of the IEEE/ACM International Conference on Automated Software Engineering*, ASE'10, pages 83–92, ACM, New York, NY, USA, 2010.

[47] A. Corradi, F. Zambonelli, and L. Leonardi. A scalable tuple space model for structured parallel programming. *Programming Models for Massively Parallel Computers, 1995*, pages 25–32, 1995.

[48] Eric Dashofy, Hazel Asuncion, Scott Hendrickson, Girish Suryanarayana, John Georgas, and Richard Taylor. Archstudio 4: An architecture-based meta-modeling environment. In *International Conference on Software Engineering*, pages 67–68. IEEE Computer Society Washington, DC, USA, 2007.

[49] Alexandre David, DeHui Du, Kim G Larsen, Marius Mikučionis, and Arne Skou. An evaluation framework for energy aware buildings using statistical model checking. *Science China Information Sciences*, 55(12):2694–2707, 2012.

[50] Alexandre David, Kim G Larsen, Axel Legay, Marius Mikučionis, and Danny Bøgsted Poulsen. Uppaal smc tutorial. *International Journal on Software Tools for Technology Transfer*, pages 1–19, 2015.

[51] Juan de Lara, Roswitha Bardohl, Hartmut Ehrig, Karsten Ehrig, Ulrike Prange, and Gabriele Taentzer. Attributed graph transformation with node type inheritance. *Theoretical Computer Science*, 376(3):139–163, 2007.

[52] Rogério De Lemos, Holger Giese, Hausi A Müller, Mary Shaw, Jesper Andersson, Marin Litoiu, Bradley Schmerl, Gabriel Tamura, Norha M Villegas, Thomas Vogel, et al. Software engineering for self-adaptive systems: A second research roadmap. In *Software Engineering for Self-Adaptive Systems II*, pages 1–32. Springer, 2013.

[53] Alan Dearle, Graham Kirby, Ron Morrison, Andrew McCarthy, Kevin Mullen, Yanyan Yang, Richard Connor, Paula Welen, and Andy Wilson. Architectural support for global smart spaces. *Lecture Notes in Computer Science*, pages 153–164, 2003.

[54] Frank DeRemer and Hans Kron. Programming-in-the-large versus programming-in-the-small. In *Proceedings of the International Conference on Reliable Software*, pages 114–121, ACM, New York, NY, USA, 1975.

[55] Anind Dey. *Providing Architectural Support for Building Context-Aware Applications*. PhD thesis, Georgia Institute of Technology, 2000.

[56] Pavlin Dobrev, David Famolari, Christian Kurzke, and Brent Miller. Device and service discovery in home networks with OSGi. *IEEE Communications Magazine*, 40(8):86–92, 2002.

[57] Jim Dowling and Vinny Cahill. Self-managed decentralised systems using K-components and collaborative reinforcement learning. In *Proceedings of the 1st ACM SIGSOFT Workshop on Self-Managed Systems*, pages 39–43. ACM, New York, NY, USA, 2004.

[58] Hartmut Ehrig, Reiko Heckel, Martin Korff, Michael Löwe, Leila Ribeiro, Annika Wagner, and Andrea Corradini. *Algebraic Approaches to Graph Transformation: Part II: Single Pushout Approach and Comparison with Double Pushout Approach*. Citeseer, 1996.

[59] Ahmed Elkhodary, Naeem Esfahani, and Sam Malek. FUSION: A framework for engineering self-tuning self-adaptive software systems. In *Proceedings of the Eighteenth ACM SIGSOFT International Symposium on Foundations of Software Engineering*, FSE'10, pages 7–16, ACM, New York, NY, USA, 2010.

[60] Tzilla Elrad, Mehmet Aksit, Gregor Kiczales, Karl Lieberherr, and Harold Ossher. Discussing aspects of AOP. *Communications of the ACM*, 44(10):33–38, 2001.

[61] Tzilla Elrad, Robert Filman, and Atef Bader. Aspect-oriented programming: Introduction. *Communications of the ACM*, 44(10):28–32, 2001.

[62] Ilenia Epifani, Carlo Ghezzi, Raffaela Mirandola, and Giordano Tamburrelli. Model evolution by run-time parameter adaptation. In *Software Engineering, 2009. ICSE 2009. IEEE 31st International Conference on*, pages 111–121. IEEE, 2009.

[63] Naeem Esfahani, Ehsan Kouroshfar, and Sam Malek. Taming uncertainty in self-adaptive software. In *Proceedings of the 19th ACM SIGSOFT*

Symposium and the 13th European Conference on Foundations of Software Engineering, ESEC/FSE'11, pages 234–244, ACM, New York, NY, USA, 2011.

[64] Naeem Esfahani and Sam Malek. Uncertainty in self-adaptive software systems. In Rogério de Lemos, Holger Giese, HausiA. Müller, and Mary Shaw, editors, *Software Engineering for Self-Adaptive Systems II*, volume 7475 of *Lecture Notes in Computer Science*, pages 214–238. Springer, Berlin, Heidelberg, 2013.

[65] Antonio Filieri, Carlo Ghezzi, and Giordano Tamburrelli. Run-time efficient probabilistic model checking. In *Proceedings of the 33rd International Conference on Software Engineering*, pages 341–350. ACM, 2011.

[66] Antonio Filieri, Lars Grunske, and Alberto Leva. Lightweight adaptive filtering for efficient learning and updating of probabilistic models. In *Proceedings of the 37th International Conference on Software Engineering*, ICSE'15. IEEE, 2015.

[67] Antonio Filieri, Henry Hoffmann, and Martina Maggio. Automated design of self-adaptive software with control-theoretical formal guarantees. In *Proceedings of the 36th International Conference on Software Engineering*, pages 299–310. ACM, 2014.

[68] Antonio Filieri, Henry Hoffmann, and Martina Maggio. Automated multi-objective control for self-adaptive software design. In *Proceedings of the 10th Joint Meeting on Foundations of Software Engineering, ESEC/FSE'15*, pages 13–24, 2015.

[69] Antonio Filieri, Giordano Tamburrelli, and Carlo Ghezzi. Supporting self-adaptation via quantitative verification and sensitivity analysis at run time. *IEEE Transactions on Software Engineering*, 42(1):75–99, 2016.

[70] Vojtech Forejt, Marta Z. Kwiatkowska, Gethin Norman, and David Parker. Automated verification techniques for probabilistic systems. In Marco Bernardo and Valérie Issarny, editors, *SFM*, volume 6659 of *Lecture Notes in Computer Science*, pages 53–113. Springer, 2011.

[71] Ian Foster, Yong Zhao, Ioan Raicu, and Shiyong Lu. Cloud computing and grid computing 360-degree compared. In *Grid Computing Environments Workshop, 2008. GCE'08*, pages 1–10, 2008.

[72] Xiang Fu, Tevfik Bultan, and Jianwen Su. Analysis of interacting BPEL web services. In *Proceedings of the 13th International Conference on World Wide Web*, pages 621–630. ACM, New York, NY, USA, 2004.

[73] Erich Gamma, Richard Helm, Ralph Johnson, and John Vlissides. *Design Patterns: Elements of Reusable Object-Oriented Software*. Addison-wesley Reading, MA, 1995.

[74] Alan Ganek and Thomas Corbi. The dawning of the autonomic computing era. *IBM Systems Journal*, 42(1):5–18, 2003.

[75] David Garlan, Shang-Wen Cheng, An-Cheng Huang, Bradley Schmerl, and Peter Steenkiste. Rainbow: Architecture-based self-adaptation with reusable infrastructure. *Computer*, 37(10):46–54, 2004.

[76] David Garlan, Robert Monroe, and David Wile. Acme: Architectural description of component-based systems. *Foundations of Component-Based Systems*, 68:47–68, 2000.

[77] David Garlan, Bradley Schmerl, and Jichuan Chang. Using gauges for architecture-based monitoring and adaptation. *Information and Software Technology*, 43, 2001.

[78] David Garlan, Daniel P. Siewiorek, Asim Smailagic, and Peter Steenkiste. Project aura: Toward distraction-free pervasive computing. *IEEE Pervasive Computing*, 1(2):22–31, 2002.

[79] John Georgas and Richard Taylor. Towards a knowledge-based approach to architectural adaptation management. In *Proceedings of the 1st ACM SIGSOFT Workshop on Self-Managed Systems*, pages 59–63. ACM, New York, NY, USA, 2004.

[80] Carlo Ghezzi, Leandro Sales Pinto, Paola Spoletini, and Giordano Tamburrelli. Managing non-functional uncertainty via model-driven adaptivity. In *Proceedings of the 2013 International Conference on Software Engineering*, ICSE'13, pages 33–42. IEEE Press, 2013.

[81] Debanjan Ghosh, Raj Sharman, Raghav Rao, and Shambhu Upadhyaya. Self-healing systems, survey and synthesis. *Decision Support Systems*, 42(4):2164–2185, 2007.

[82] Y.Y. Goland, T. Cai, P. Leach, Y. Gu, and S. Albright. Simple service discovery protocol/1.0 operating without an arbiter. *Internet Engineering Task Force, Draft, October*, 188, 1999.

[83] P. Grace, G.S. Blair, and S. Samuel. A reflective framework for discovery and interaction in heterogeneous mobile environments. *ACM SIGMOBILE Mobile Computing and Communications Review*, 9(1):2–14, 2005.

[84] Robert Grimm et al. One.world: Experiences with a pervasive computing architecture. *IEEE Pervasive Computing*, 3(3):22–30, 2004.

[85] T.R. Gruber. A translation approach to portable ontology specifications. *Knowledge Acquisition*, 5(2):199–220, 1993.

[86] Tao Gu, Xiao Hang Wang, Hung Keng Pung, and Da Qing Zhang. An ontology-based context model in intelligent environments. In *Proceedings of Communication Networks and Distributed Systems Modeling and Simulation Conference*, volume 2004, 2004.

[87] Erik Guttman and John Veizades. Service Location Protocol, Version 2. *IETF RFC 2165*, 1999.

[88] Annegret Habel, Reiko Heckel, and Gabriele Taentzer. Graph grammars with negative application conditions. *Fundamenta Informaticae*, 26(3-4):287–313, 1996.

[89] Arnd Hartmanns and Holger Hermanns. A modest approach to checking probabilistic timed automata. In *Quantitative Evaluation of Systems, 2009. QEST'09. Sixth International Conference on the*, pages 187–196. IEEE, 2009.

[90] Jan Hendrik Hausmann, Reiko Heckel, and Gabi Taentzer. Detection of conflicting functional requirements in a use case-driven approach: a static analysis technique based on graph transformation. In *Proceedings of the 24th International Conference on Software Engineering*, pages 105–115. ACM, New York, NY, USA, 2002.

[91] Christopher Hayden, Stephen Magill, Michael Hicks, Nate Foster, and Jeffrey Foster. Specifying and verifying the correctness of dynamic software updates. In *Verified Software: Theories, Tools, Experiments*, pages 278–293. Springer, 2012.

[92] Brian Hayes. Cloud computing. *Commun. ACM*, 51(7), 2008.

[93] Reiko Heckel and Annika Wagner. Ensuring consistency of conditional graph grammars-a constructive approach. *Electronic Notes in Theoretical Computer Science*, 2:118–126, 1995.

[94] Albert Held, Sven Buchholz, and Alexander Schill. Modeling of context information for pervasive computing applications. In *Proceedings of the 6th World Multiconference on Systemics, Cybernetics and Informatics (SCI)*, 2002.

[95] Karen Henricksen, Jadwiga Indulska, and Andry Rakotonirainy. Generating context management infrastructure from high-level context models. In *Proceedings of the 4th International Conference on Mobile Data Management*, pages 1–6, 2003.

[96] Karsten Hölscher, Paul Ziemann, and Martin Gogolla. On translating UML models into graph transformation systems. *Journal of Visual Languages & Computing*, 17(1):78–105, 2006.

[97] Ian Horrocks. DAML+ OIL: a reason-able web ontology language. In *Proceedings of the 8th International Conference on Extending Database Technology*, page 2. Springer, 2002.

[98] Gang Huang, Hong Mei, and Fuqing Yang. Runtime software architecture based on reflective middleware. *Science in China Series F: Information Sciences*, 47(5):555–576, 2004.

[99] Michael Huhns, Vance Holderfield, and Rosa Laura Zavala Gutierrez. Robust software via agent-based redundancy. In *Proceedings of the Second International Joint Conference on Autonomous Agents and Multiagent Systems*, pages 1018–1019. ACM, New York, NY, USA, 2003.

[100] Michael Huhns and Munindar Singh. Service-oriented computing: Key concepts and principles. *IEEE Internet Computing*, 9(1):75–81, 2005.

[101] IBM. An architectural blueprint for autonomic computing. 2006.

[102] B. Jacob, International Technical Support Organization, and International Business Machines Corporation. *A Practical Guide to the IBM Autonomic Computing Toolkit*. IBM, International Technical Support Organization, 2004.

[103] Sae Hoon Kang, Seungbok Ryu, Namhoon Kim, Younghee Lee, Dongman Lee, and Keyong-Deok Moon. An architecture for interoperability of service discovery protocols using dynamic service proxies. In *ICOIN*, pages 786–795. Springer, 2005.

[104] Kamran Karimi, Neil Dickson, and Firas Hamze. High-performance physics simulations using multi-core cpus and gpgpus in a volunteer computing context. *International Journal of High Performance Computing Applications*, 25(1):61–69, 2011.

[105] Raman Kazhamiakin, Paritosh Pandya, and Marco Pistore. Timed modelling and analysis in web service compositions. In *Availability, Reliability and Security, 2006. ARES 2006. The First International Conference on*, pages 7–pp. IEEE, 2006.

[106] Jeffrey Kephart and David Chess. The vision of autonomic computing. *Computer*, 36(1):41–50, 2003.

[107] Gregor Kiczales, Erik Hilsdale, Jim Hugunin, Mik Kersten, Jeffrey Palm, and William Griswold. An overview of AspectJ. *Lecture Notes in Computer Science*, pages 327–353, 2001.

[108] Teemu Koponen and Teemupekka Virtanen. A service discovery: A service broker approach. In *System Sciences, 2004. Proceedings of the 37th Annual Hawaii International Conference on*, page 7, 2004.

[109] Jeff Kramer and Jeff Magee. The evolving philosophers problem: Dynamic change management. *Software Engineering, IEEE Transactions on*, 16(11):1293–1306, 1990.

[110] Jeff Kramer and Jeff Magee. Self-managed systems: an architectural challenge. In *International Conference on Software Engineering*, pages 259–268. IEEE Computer Society Washington, DC, USA, 2007.

[111] Philippe Kruchten. The 4 + 1 view model of architecture. *IEEE Software*, 12(6):42–50, 1995.

[112] Naveen Kumar, Jonathan Misurda, Bruce Childers, and Mary Lou Soffa. Instrumentation in software dynamic translators for self-managed Systems. In *Proceedings of the 1st ACM SIGSOFT Workshop on Self-Managed Systems*, pages 90–94. ACM, New York, NY, USA, 2004.

[113] Leen Lambers, Hartmut Ehrig, and Fernando Orejas. Conflict detection for graph transformation with negative application conditions. *Lecture Notes in Computer Science*, 4178:61, 2006.

[114] D.B. Lange and M. Oshima. Seven good reasons for mobile agents. *Communications of the ACM*, 42(3):88–89, 1999.

[115] Daniel Le Métayer. Describing software architecture styles using graph grammars. *Software Engineering, IEEE Transactions on*, 24(7):521–533, 1998.

[116] Axel Legay, Benoît Delahaye, and Saddek Bensalem. Statistical model checking: An overview. In *Runtime Verification*, pages 122–135. Springer, 2010.

[117] Emmanuel Letier, David Stefan, and Earl T. Barr. Uncertainty, risk, and information value in software requirements and architecture. In *Proceedings of the 36th International Conference on Software Engineering*, ICSE'14, pages 883–894, ACM, New York, NY, USA, 2014.

[118] K. Lieberherr. Workshop on adaptable and adaptive software. *ACM SIGPLAN OOPS Messenger*, 6(4):149–154, 1995.

[119] Jian Lü, Xiaoxing Ma, XianPing Tao, Chun Cao, Yu Huang, and Ping Yu. On environment-driven software model for internetware. *Science in China Series F: Information Sciences*, 51(6):683–721, 2008.

[120] Jian Lü, Xiaoxing Ma, Xianping Tao, Feng Xu, and Hao Hu. Research and progress on internetware. *Science in China (Series E)*, 36(10):1037–1080, 2006.

[121] Xiaoxing Ma, Yu Zhou, Jian Pan, Ping Yu, and Jian Lu. Constructing self-adaptive systems with polymorphic software architecture. In *Proceedings of the 19th Int'l Conf. on Software Engineering and Knowledge Engineering*. Knowledge Systems Institute Graduate School, 2007.

[122] Jeff Magee and Jeff Kramer. Dynamic structure in software architectures. In *Foundations of Software Engineering: Proceedings of the 4th ACM SIGSOFT Symposium on Foundations of Software Engineering*. ACM, New York, USA, 1996.

[123] Brian McBride. Jena: A semantic web toolkit. *IEEE Internet Computing*, 6(6):55–59, 2002.

[124] John McCarthy. Notes on formalizing context. In *International Joint Conference on Artificial Intelligence*, volume 13, pages 555–555. Lawrence Erlbaum Associates Ltd., 1993.

[125] Nenad Medvidovic, Eric Dashofy, and Richard Taylor. Moving architectural description from under the technology lamppost. *Information and Software Technology*, 49(1):12–31, 2007.

[126] Nenad Medvidovic and Richard Taylor. A classification and comparison framework for software architecture description languages. *IEEE Transactions on Software Engineering*, 26(1):70–93, 2000.

[127] Hong Mei, Gang Huang, Lu Zhang, and Wei Zhang. ABC: a method of software architecture modeling in the whole lifecycle. *Science in China Series F-Information Sciences (in Chinese)*, 44(5):564–587, 2014.

[128] Peter Mell and Tim Grance. *The NIST Definition of Cloud Computing*. Computer Security Division, Information Technology Laboratory, National Institute of Standards and Technology, Gaithersburg, MD, 2011.

[129] Tom Mens, Gabriele Taentzer, and Olga Runge. Detecting structural refactoring conflicts using critical pair analysis. *Electronic Notes in Theoretical Computer Science*, 127(3):113–128, 2005.

[130] Tom Mens, Gabriele Taentzer, and Olga Runge. Analysing refactoring dependencies using graph transformation. *Software and Systems Modeling*, 6(3):269–285, 2007.

[131] Tom Mens, Ragnhild Van Der Straeten, and Maja DHondt. Detecting and resolving model inconsistencies using transformation dependency analysis. *Lecture Notes in Computer Science*, 4199:200, 2006.

[132] Robin Milner. *Communicating and Mobile Systems: the Pi Calculus.* Cambridge University Press, 1999.

[133] Gabriel Moreno, Javier Cámara, David Garlan, and Bradley Schmerl. Proactive self-adaptation under uncertainty: A probabilistic model checking approach. In *Proceedings of the 10th Joint Meeting on Foundations of Software Engineering*, ESEC/FSE 2015, pages 1–12, ACM, New York, NY, USA, 2015.

[134] OMG. Specification. unified modeling language: Superstructure version 2.2. *OMG Formal Document*, 2009.

[135] Flavio Oquendo, Brian Warboys, Ron Morrison, Régis Dindeleux, Ferdinando Gallo, Hubert Garavel, and Carmen Occhipinti. Archware: Architecting evolvable software. *Lecture Notes in Computer Science*, pages 257–271, 2004.

[136] Peyman Oreizy, Nenad Medvidovic, and Richard N. Taylor. Architecture-based runtime software evolution. In *ICSE '98: Proceedings of the 20th International Conference on Software Engineering*, pages 177–186, Washington, DC, USA, 1998. IEEE Computer Society.

[137] Peyman Oreizy, Nenad Medvidovic, and Richard N. Taylor. Runtime software adaptation: framework, approaches, and styles. In *ICSE Companion '08*, ACM, New York, NY, USA, 2008.

[138] Dimitri Papadimitriou et al. Future internet–the cross-etp vision document. *European Technology Platform, Alcatel Lucent*, 8:3, 2009.

[139] George Papadopoulos and Farhad Arbab. Modelling activities in information systems using the coordination language Manifold. In *Proceedings of the 1998 ACM Symposium on Applied Computing*, pages 185–193. ACM, New York, NY, USA, 1998.

[140] Mike Papazoglou. Service-oriented computing: Concepts, characteristics and directions. In *Web Information Systems Engineering, 2003. WISE 2003. Proceedings of the Fourth International Conference on*, pages 3–12, 2003.

[141] David Lorge Parnas. On the criteria to be used in decomposing systems into modules. *Communications of the ACM*, 15(12):1053–1058, 1972.

[142] Chris Peltz. Web services orchestration and choreography. *Computer*, 36(10):46–52, 2003.

[143] Detlef Plump. Hypergraph rewriting: Critical pairs and undecidability of confluence. *Term Graph Rewriting: Theory and Practice*, 1993.

[144] Vahe Poladian, David Garlan, Mary Shaw, Mahadev Satyanarayanan, Bradley Schmerl, and Joao Sousa. Leveraging resource prediction for anticipatory dynamic configuration. In *Self-Adaptive and Self-Organizing Systems, 2007. SASO '07. First International Conference on*, pages 214–223, 2007.

[145] E. PrudHommeaux, A. Seaborne, et al. SPARQL query language for RDF. *W3C working draft*, 4, 2006.

[146] Martin Puterman. Markov decision processes. *Handbooks in Operations Research and Management Science*, 2:331–434, 1990.

[147] Anand Ranganathan, Shiva Chetan, and Roy Campbell. Mobile polymorphic applications in ubiquitous computing environments. In *Mobile and Ubiquitous Systems: Networking and Services, 2004. Mobiquitous 2004. The First Annual International Conference on*, pages 402–411, 2004.

[148] Awais Rashid and Gerd Kortuem. Adaptation as an aspect in pervasive computing. In *Workshop on Building Software for Pervasive Computing at OOPSLA*, 2004.

[149] Pierre-Guillaume Raverdy, Valerie Issarny, Rafik Chibout, and Agnes de La Chapelle. A multi-protocol approach to service discovery and access in pervasive environments. In *Proceedings of Mobiquitous - The 3rd Annual International Conference on Mobile and Ubiquitous Systems: Networks and Services*, 2006.

[150] Golden III Richard. Service advertisement and discovery: Enabling universal device cooperation. *IEEE Internet Computing*, 4(5):18–26, 2000.

[151] Matthias Rohr, Simon Giesecke, Marcel Hiel, Willem-Jan van den Heuvel, Hans Weigand, and Wilhelm Hasselbring. A classification scheme for self adaptation research. In *International Conference on Self-Organization and Autonomous Systems in Computing and Communications (SOAS06) Poster Session, Erfurt, Germany*, 2006.

[152] G.C. Roman, C. Julien, and J. Payton. Modeling adaptive behaviors in Context UNITY. *Theoretical Computer Science*, 376(3):185–204, 2007.

[153] Manuel Román, Christopher Hess, Renato Cerqueira, Anand Ranganat, Roy Campbell, and Klara Nahrstedt. Gaia: A middleware infrastructure to enable active spaces. *IEEE Pervasive Computing*, 1(4):74–83, 2002.

[154] Grzegorz Rozenberg. *Handbook of Graph Grammars and Computing by Graph Transformation: Volume I. Foundations*. World Scientific Publishing Co., Inc., River Edge, NJ, USA, 1997.

[155] Larry Rudolph. Project oxygen: Pervasive, human-centric computing-an initial experience. *Lecture Notes in Computer Science*, pages 1–12, 2001.

[156] Nick Ryan, Jason Pascoe, and David Morse. Enhanced reality fieldwork: the context aware archaeological assistant. In *Archaeology in the Age of the Internet: CAA 97: Computer Applications and Quantitative Methods in Archaeology: Proceedings of the 25th Anniversary Conference*, page 269. British Archaeological Reports, 1999.

[157] Debashis Saha and Amitava Mukherjee. Pervasive computing: A paradigm for the 21st century. *Computer*, 36(3):25–31, 2003.

[158] Mazeiar Salehie and Ladan Tahvildari. Self-adaptive software: Landscape and research challenges. *ACM Trans. Auton. Adapt. Syst.*, 4(2):14:1–14:42, May 2009.

[159] Mahadev Satyanarayanan. Pervasive computing: Vision and challenges. *IEEE [see also IEEE Wireless Communications] Personal Communications*, 8(4):10–17, 2001.

[160] Albrecht Schmidt, Michael Beigl, and Hans-W. Gellersen. There is more to context than location. *Computers & Graphics*, 23(6):893–901, 1999.

[161] Steve Shafer, John Krumm, Barry Brumitt, Brian Meyers, Mary Czerwinski, and Daniel Robbins. The new easy living project at Microsoft Research. In *Proceedings of the 1998 DARPA/NIST Smart Spaces Workshop*, pages 127–130, 1998.

[162] Mary Shaw and David Garlan. *Software Architecture: Perspectives on an Emerging Discipline*. Prentice-Hall, Inc., Upper Saddle River, NJ, USA, 1996.

[163] Jamie Shiers. The worldwide lhc computing grid (worldwide lcg). *Computer Physics Communications*, 177(1):219–223, 2007.

[164] Wei Song, JinHui Tang, GongXuan Zhang, and XiaoXing Ma. Substitutability analysis of ws-bpel services. *China Science: Information Science*, 42(3):264–279, 2012.

[165] Nikolaos Spanoudakis and Pavlos Moraitis. Engineering ambient intelligence systems using agent technology. *IEEE Intelligent Systems*, 30(3):60–67, 2015.

[166] Thomas Strang and Claudia Linnhoff-Popien. A context modeling survey. In *Workshop on Advanced Context Modelling, Reasoning and Management as Part of UbiComp*, 2004.

[167] Thomas Strang, Claudia Linnhoff-Popien, and Korbinian Frank. Applications of a context ontology language. In *Proceedings of International Conference on Software, Telecommunications and Computer Networks (SoftCom2003)*, volume 14, page 18, 2003.

[168] Sylvia Stuurman and Jan Van Katwijk. On-line change mechanisms: The software architectural level. In *Proceedings of the 6th ACM SIGSOFT International Symposium on Foundations of Software Engineering*, pages 80–86. ACM New York, NY, USA, 1998.

[169] Guoxin Su, Taolue Chen, Yuan Feng, David S. Rosenblum, and P. S. Thiagarajan. An iterative decision-making scheme for markov decision processes and its application to self-adaptive systems. In Perdita Stevens and Andrzej Wasowski, editors, *Fundamental Approaches to Software Engineering - 19th International Conference, FASE 2016, Held as Part of the European Joint Conferences on Theory and Practice of Software, ETAPS 2016, Eindhoven, The Netherlands, April 2-8, 2016, Proceedings*, volume 9633 of *Lecture Notes in Computer Science*, pages 269–286. Springer, 2016.

[170] Narayanan Subramanian. *Adaptable Software Architecture Generation Using the NFR Approach*. PhD thesis, 2003.

[171] Gabriele Taentzer. AGG: A graph transformation environment for modeling and validation of software. *Lecture Notes in Computer Science*, 3062:446–453, 2004.

[172] Peter Tandler. The beach application model and software framework for synchronous collaboration in ubiquitous computing environments. *Journal of Systems and Software*, 69(3):267–296, 2004.

[173] XianPing Tao, XiaoXing Ma, Jian Lu, Ping Yu, and Yu Zhou. Multi-mode interaction middleware for software services. *Science in China Series F: Information Sciences*, 51(8):985–999, 2008.

[174] Richard Taylor, Nenad Medvidovic, and Eric Dashofy. *Software Architecture: Foundations, Theory, and Practice*. Wiley Publishing, 2009.

[175] Richard Taylor and Andre van der Hoek. Software design and architecture: The once and future focus of software engineering. In *FOSE '07: 2007 Future of Software Engineering*, pages 226–243, Washington, DC, USA, 2007. IEEE Computer Society.

[176] Mike Uschold and Michael Gruninger. Ontologies: Principles, methods and applications. *Knowledge Engineering Review*, 1996.

[177] Giuseppe Valetto and Gail Kaiser. Using process technology to control and coordinate software adaptation. In *ICSE '03: Proceedings of the 25th International Conference on Software Engineering*, pages 262–272, Washington, DC, USA, 2003. IEEE Computer Society.

[178] Yves Vandewoude, Peter Ebraert, Yolande Berbers, and Theo D'Hondt. Tranquility: A low disruptive alternative to quiescence for ensuring safe dynamic updates. *Software Engineering, IEEE Transactions on*, 33(12):856–868, 2007.

[179] Richard Varga. *Matrix Iterative Analysis (Springer Series in Computational Mathematics)*. Springer, 2009.

[180] Mladen Vouk. Cloud computing–issues, research and implementations. *Journal of Computing and Information Technology*, 16(4), 2004.

[181] Jim Waldo, Ken Arnold, et al. *The Jini Specifications*. Addison-Wesley, Longman Publishing Co., Inc., Boston, MA, USA, 2000.

[182] Qianxiang Wang, Junrong Shen, Xiaopeng Wang, and Hong Mei. A component-based approach to online software evolution. *Journal of Software Maintenance and Evolution: Research and Practice*, 18(3), 2006.

[183] Xiao Hang Wang, Daqing Zhang, Tao Gu, and Hung Keng Pung. Ontology based context modeling and reasoning using OWL. In *Pervasive Computing and Communications Workshops, 2004. Proceedings of the Second IEEE Annual Conference on*, pages 18–22, 2004.

[184] Michel Wermelinger and José Luiz Fiadeiro. A graph transformation approach to software architecture reconfiguration. *Science of Computer Programming*, 44(2):133–155, 2002.

[185] Chang Xu, YePang Liu, Shing Chi Cheung, Chun Cao, and Jian Lv. Towards context consistency by concurrent checking for internetware applications. *Science China Information Sciences*, 56(8):1–20, 2013.

[186] Hongzhen Xu, Guosun Zeng, and Bo Chen. Conditional hypergraph grammars and its analysis of dynamic evolution of software architectures. *Chinese Journal of Software*, 22(6):1210–1223, 2011.

[187] Fuqing Yang, Hong Mei, Jian Lu, and Zhi Jin. Some discussion on the development of software technology. *Acta Electronica Sinica*, 30(12A):1901–1906, 2002.

[188] Juan Ye, Lorcan Coyle, Simon Dobson, and Paddy Nixon. Ontology-based models in pervasive computing systems. *The Knowledge Engineering Review*, 22(04):315–347, 2007.

[189] Ping Yu, Jiannong Cao, Weidong Wen, and Jian Lu. Mobile agent enabled application mobility for pervasive computing. *Lecture Notes in Computer Science*, 4159:648, 2006.

[190] Ping Yu, Xiaoxing Ma, Jiannong Cao, and Jian Lu. Application mobility in pervasive computing: A survey. *Pervasive and Mobile Computing*, 9(1):2–17, 2013.

[191] Jin Zeng, Hai-Long Sun, Xu-Dong Liu, Ting Deng, and Jin-Peng Huai. Dynamic evolution mechanism for trustworthy software based on service composition. *Journal of Software*, 21(2):261–276, 2010.

[192] Ji Zhang and Betty Cheng. Model-based development of dynamically adaptive software. In *Proceedings of the 28th International Conference on Software Engineering*, ICSE'06, pages 371–380, ACM, New York, NY, USA, 2006.

[193] Pengcheng Zhang, Hareton Leung, Wenrui Li, and Xuandong Li. Web services property sequence chart monitor: a tool chain for monitoring bpel-based web service composition with scenario-based specifications. *IET Software*, 7(4):222–248, 2013.

[194] Yu Zhou. A runtime architecture-based approach for the dynamic evolution of distributed component-based systems. In *ICSE Companion '08*, ACM, New York, NY, USA, 2008.

[195] Yu Zhou, Luciano Baresi, and Matteo Rossi. Towards a formal semantics for uml/marte state machines based on hierarchical timed automata. *Journal of Computer Science and Technology*, 28(1):188–202, 2013.

[196] Yu Zhou, Xiaoxing Ma, and Harald Gall. A middleware platform for the dynamic evolution of distributed component-based systems. *Computing*, 96(8):725–747, 2014.

[197] Yu Zhou, Jian Pan, Xiaoxing Ma, Bin Luo, Xianping Tao, and Jian Lu. Applying ontology in architecture-based self-management applications. In *Proceedings of the 2007 ACM Symposium on Applied Computing*, pages 97–103. ACM, New York, NY, USA, 2007.

[198] Yu Zhou, Xuefeng Yan, and Zhiqiu Huang. A graph transformation based approach for modeling component-level migration in mobile environments. In *Computer Software and Applications Conference Workshops (COMPSACW), 2012 IEEE 36th Annual*, pages 152–157. IEEE, 2012.

[199] Fen Zhu, Matt Mutka, and Lionel Ni. Service discovery in pervasive computing environments. *IEEE Pervasive Computing*, 4(4):81–90, 2005.

Index

Printed and bound by CPI Group (UK) Ltd, Croydon, CR0 4YY

24/10/2024

01778284-0004